TH   Slater, H.
6123
S53   Advanced plumbing.
1980

Y

Advanced Plumbing

THIRD EDITION

Harry Slater & Lee Smith

Advanced Plumbing

THIRD EDITION

Harry Slater & Lee Smith

VNR

VAN NOSTRAND REINHOLD COMPANY

NEW YORK CINCINNATI TORONTO LONDON MELBOURNE

Copyright © 1980 by Litton Educational Publishing, Inc.
Library of Congress Catalog Card Number 79-27408
ISBN 0-442-23873-8

Printed in the United States of America.

Published by Van Nostrand Reinhold Company
A division of Litton Educational Publishing, Inc.
135 West 50th Street, New York, NY 10020, U.S.A.

Van Nostrand Reinhold Limited
1410 Birchmount Road
Scarborough, Ontario M1P 2E7, Canada

Van Nostrand Reinhold Australia Pty. Ltd.
17 Queen Street
Mitcham, Victoria 3132, Australia

Van Nostrand Reinhold Company Limited
Molly Millars Lane
Wokingham, Berkshire, England

16 15 14 13 12 11 10 9 8 7 6 5 4 3 2 1

Library of Congress Cataloging in Publication Data

Slater, Harry.
  Advanced plumbing.

    Revision of v. 2 of the 1974-75 ed. of the author's
Plumbing.
    1. Plumbing.   I. Smith, Lee, joint author.
II. Title.
TH6123.S53 1980   696'.1   79-27408
ISBN 0-442-23873-8

Preface

Plumbing as a trade grows in importance as our society grows in complexity. It is an occupation vital to the health and well-being of society. Today's plumber must keep abreast of a changing technology in order to install and repair a variety of plumbing systems.

ADVANCED PLUMBING builds on basic plumbing knowledge and skills. It combines up-to-date information with sound teaching methods to prepare plumbers for their roles as skilled craftspeople.

This revision of ADVANCED PLUMBING reflects current practices in the plumbing industry. The units on Plastic Pipe and Fittings and Solar Water Heating have been expanded to include more comprehensive information. New units include Welded Pipe, Private Water Systems, Sizing the Water Supply System and Fixture Installation. The text also introduces the SOVENT drainage system.

The units in ADVANCED PLUMBING are arranged in a logical order of study. Each unit begins with a set of objectives. These clearly state what you will learn in the unit. The unit-end review questions give you a chance to apply the information presented in the unit.

Lee Smith, the revising author, has been an instructor of plumbing in vocational high schools, in-service training programs in industry, and Manpower, Inc. He is a master plumber with many years experience.

Contents

SECTION 1

Pipe and Fittings

UNIT 1 PLASTIC PIPE AND FITTINGS

OBJECTIVES

After studying this unit, the student will be able to:

- describe the differences between plastic pipe materials.
- assemble different kinds of plastic pipe.
- explain the uses of plastic pipe materials.

PLASTIC PIPE AND FITTINGS

Plastic refers to a family of chemical compositions. While these chemical compositions may look the same, their uses in a plumbing system may be very different.

New plastic pipe materials are being introduced every day. Many plumbing codes now permit the use of plastic pipe. Some companies are guaranteeing their plastic pipe and fittings for a fifty-year period.

Plastic piping has certain advantages and disadvantages when compared to metal piping.

Advantages:

- It is lightweight.
- It is less expensive.
- It resists corrosion.
- It is fast and easy to install.
- It can be made with special characteristics.

Disadvantages:

- It is less resistant to heat.
- It expands and contracts more when heated and cooled.
- It needs more support because of its flexibility.
- It has less crush resistance.
- It withstands less internal pressure.

Three common types of plastic pipe and fittings used in plumbing are *ABS* (acrylonitrile butadiene styrene), *PE* (polyethylene), and *PVC* and *CPVC* (polyvinyl chloride and chlorinated polyvinyl chloride). All have different applications in the plumbing system.

ABS PIPING

ABS piping is used extensively for drainage, waste, and vent piping. While ABS withstands heat well, a 10-foot section of steel pipe is recommended on commercial dishwasher drains. This dispels the heat before the waste water enters the plastic.

ABS pipe does not thread satisfactorily. An adapter is necessary to connect it to a threaded outlet. Horizontal piping must be supported at 4-foot intervals or less to avoid sagging. A lead-poured joint may be used to join plastic to a cast-iron bell. The lead will cool and solidify before the plastic begins to melt.

When connecting threaded joints, petroleum jelly or Teflon® tape is used for pipe dope. A strap wrench is used for assembling threaded joints. Pipe wrenches scar the pipe fitting and may crack it or cause a weak spot. The tools necessary for roughing-in with plastic pipe include a rule, a tubing cutter with a special wheel, a half-round file for deburring, and a narrow paint brush for applying cement. A piece of sandpaper for cleaning dirty ABS and PVC pipe is also helpful.

ABS End Preparation (Solvent Weld)

1. Cut the pipe with a fine-tooth saw in a miter box, or with a tubing cutter with a special wheel, figure 1-1.

2. Deburr the pipe inside and out with a half-round file, figure 1-2.

3. Clean the pipe end of dirt, paint, or grease.

4. Apply solvent cement to both pipe and fitting, figure 1-3.

5. Quickly insert the pipe all the way into the fitting. Twist the pipe 1/4 turn to evenly coat the joint, figure 1-4. There should be a full bead around the socket with no gaps. Wipe off any excess cement, leaving a small fillet between the pipe and fitting.

6. Wait 3 minutes before handling.

The pipe will be ready to water test within 2 hours. Because the cement dries so

Fig. 1-1 Cutting pipe with a tube cutter

Fig. 1-2 Deburring with a half-round file

Fig. 1-3 Apply solvent cement to pipe and fitting.

Fig. 1-4 Insert pipe into fitting and twist.

quickly, angles between fittings must be set quickly or the joint will be impossible to turn. Plastic fittings cannot be reused without expensive machining.

PE PIPING

PE or polyethylene pipe is a flexible plastic pipe used for underground installations, such as wells, sprinkler systems, and water supply systems. It is available in long coils of 500 feet. Maximum operating pressures vary with the particular quality and compound materials of a given product, but PE pipe may be obtained with guaranteed pressures of 80 to 160 pounds per square inch (psi).

PE piping is normally used outside of the foundation walls of a house. It is not used for hot water. The piping lays in a trench that is at least 12 inches below the frost line. The trench must be free of rocks and other sharp projections where the pipe is to be laid.

All plastics have a high expansion rate compared to steel pipe. Therefore, plastic pipe should be allowed to snake or wind down the trench. Temperatures underground are relatively constant, however, and expansion problems are rare.

Backfilling the trench must be done carefully to a depth of 6 inches above the pipe. Before backfilling further, the pipe should be pressure-tested to 150 percent of its working pressure.

PE End Preparation (Insert Filling Joint)

1. Cut the pipe squarely with a saw or tubing cutter. Ream the end with a half-round file.

2. Slip two hose clamps onto the pipe end.

3. Press the pipe onto the fitting without any pipe dope. If the fit is too tight, dip the pipe end in a weak soap solution. Dipping the pipe end into very hot water will also help.

4. Tighten the hose clamps. Note that the screws are 180 degrees apart to distribute tension evenly, figure 1-5.

Polyethylene pipe may also be cold-flared with a special flaring tool. It is then assembled the same as any flared fitting. Insert fittings for polyethylene pipe are made of nylon or PVC.

PVC AND CPVC

PVS and CPVC plastic pipe materials are used for high pressure applications (up to 400 psi). The dimensions of schedule #40 PVC (standard weight) are the same as those for standard steel pipe. CPVC is also designed to handle hot liquids up to 180 degrees Fahrenheit. Both come in rigid form.

PVC is used above and below ground level in such installations as underground pipelines, factories, and multiunit housing. It is threaded in schedule #80 or heavier form. Standard pipe dies may be used. They must be sharp and only used for plastic pipe. CPVC is used in the home for water supply piping and for smaller water service piping anywhere.

Both materials are tested with the pipe exposed. Allow sufficient drying time before pressure testing. Installing one day and testing the next day is usually sufficient. Because of long setting times, PVS and CPVC require more skill and planning to install than other kinds of plastics. The pipe fitter must consider

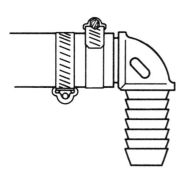

Fig. 1-5

the properties of the material and the temperature and weather conditions at the time of installation.

PVC sags at a sustained internal or external temperature of 110 degrees. To install pipe horizontally at this or higher temperatures, it must be supported along its entire length. All plastic pipe materials have a much higher expansion rate than steel pipe. CPVC expands 1/2 inch in a 10-foot length for a temperature rise between 73 and 180 degrees. Contraction occurs at the same rate for a drop in temperature. If a pipeline is installed in midsummer and fastened securely at both ends, the joints might pull apart when the temperature drops below 32 degrees the following winter.

PVC, like most plastics, must be well supported when it is stacked for storage, or it may take a permanent set. Dragging the pipe across a concrete floor will wear flat spots on the pipe and weaken it. Pipe hangers should not hold the pipe tightly. Strap hangers 3/4 inch or wider are best. Do not use wire hangers. The pipe should be free to slide in the hangers. Place hangers at every other joist or every 2 1/2 to 3 1/2 feet.

When joining PVC and CPVC to metal pipe, a special adapter, called a transition fitting, is used rather than a standard adapter. The transition fitting allows for the expansion difference between the two materials.

PVC and CPVC End Preparation

The assembly of PVC, CPVC, and ABS is similar with two exceptions. PVC and CPVC use a primer before the solvent cement is applied. They are also assembled with great care.

ABS solvent cement is not interchangeable with CPVC or PVC solvent cement. The instructions on the can must be followed exactly. After assembly, avoid disturbing the joint. With PVC, a bevel is often filed on the outside end of the pipe so that cement is not pushed inside the fitting.

REVIEW QUESTIONS

Multiple Choice

Select the best answer for each question.

1. ABS piping is used mainly for
   a. drainage, waste, and venting.
   b. water supply.
   c. flexible well piping.
   d. fittings.

2. ABS piping should be supported at
   a. 2-foot intervals.
   b. 4-foot intervals.
   c. 6 to 8-foot intervals.
   d. 20-foot intervals.

3. What type of wrench should be used for tightening plastic threaded joints?
   a. Pipe wrench
   b. Strap wrench
   c. Adjustable (crescent) wrench
   d. Basin wrench

4. When cementing ABS pipe, how long should the plumber wait before handling joints after initial assembly?
   a. 2 hours
   b. 1 hour
   c. 30 minutes
   d. 3 minutes

5. After assembly, how long should the plumber wait before water testing ABS pipe?

   a. 30 minutes   c. 12 hours

   b. 2 hours   d. 24 hours

6. Polyethylene plastic is used extensively for

   a. well piping.   c. radiant heating panels.

   b. hydraulics.   d. pressure piping.

7. The _____ of all plastic pipe materials is much greater than comparable metal pipe.

   a. corrosion rate   c. expansion rate

   b. friction loss   d. weight

8. The best polyethylene pipe will not withstand working pressures greater than

   a. 40 psi.   c. 80 psi.

   b. 60 psi.   d. 160 psi.

9. PVC plastic pipe may be threaded if

   a. a well-worn threading die is used.

   b. it is schedule #80 or heavier.

   c. it is no heavier than standard weight.

   d. plastic dies are used.

10. The plastic pipe material requiring the most care in installation is

    a. ABS.   c. PVC or CPVC.

    b. polyethylene.   d. polybutylene.

11. Hangers for plastic pipe should

    a. allow the pipe to slide through.

    b. be the wire type.

    c. restrict the movement of the pipe.

    d. be grounded.

12. What should be used to join PVC or CPVC to a high-pressure metal pipe?

    a. Adapter   c. Transition fitting

    b. Coupling   d. 90-degree elbow

UNIT 2 BITUMINIZED-FIBER SEWER PIPE

OBJECTIVES

After studying this unit, the student will be able to:

- describe the features, types, and uses of bituminized-fiber sewer pipe.
- select the fittings to use with bituminized-fiber sewer pipe.

BITUMINIZED-FIBER SEWER PIPE

Bituminized-fiber pipe is made of wood fibers saturated with coal-tar pitch and heated to 340 degrees Fahrenheit in a vacuum.

Fiber pipe is made in the following sizes.

- 2" diameter by 5' long
- 3" diameter by 8' long
- 4" diameter by 8' long
- 5" diameter by 5' long
- 6" diameter by 5' long

An 8-foot length of 4-inch fiber pipe weighs only 2.7 pounds per foot or a total of 21.6 pounds.

Each length of fiber has a factory-machined, 2-degree taper on each end. The fiber couplings and fittings have an inverted 2-degree taper into which the pipe is driven when joined together. This is called a *taperweld joint*, figure 2-1. It is swaged together without joining the components.

Fiber pipe is installed by placing a coupling or fitting on the end of a length of pipe. A block of wood is placed against the coupling and tapped lightly with a hammer. The coupling must be placed in a straight line with the pipe. A hammer is never used on the end of the pipe. This would damage the pipe. Five-degree angle couplings are used if pipe direction must be changed.

Fiber pipe may be cut with ordinary woodworking tools. The taperweld joint is usually made on the job with a field tooling lathe.

Fiber pipe is not used on pressure systems. It is only recommended for gravity flow sewer lines outside buildings. Most plumbing codes insist on the use of cast-iron pipe in streets.

Other restrictions and disadvantages to bituminized-fiber sewer pipe include the following:

- It is more expensive than plastic pipe.
- Installation is slow and difficult.
- It should not be exposed to direct sunlight.
- It is relatively weak and therefore requires the use of backfill.

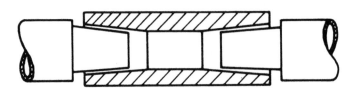

Fig. 2-1 Taperweld joint

• Check local codes before using it. Many codes specifically forbid using bituminized-fiber sewer pipe.

BITUMINIZED-FIBER FITTINGS

Fiber fittings, made by the Orangeburg Company, are generally called *Orangeburg fittings,* figure 2-2. These fittings are available in the following sizes:

• 4″ one-eighth bends

• 4″ one-quarter bends

• long sweep 45° bends

• long sweep 90° bends

Adapters are also available to attach fiber pipe to clay, soil, or screw pipe. By using adapters with wyes, a variety of sizes and connections may be obtained. For instance, a combination wye and one-eighth bend may be joined as in figure 2-3. The knobs on the elbows in the one-eighth and one-quarter bends in figure 2-2 are used to drive the fittings into the pipes when making the joint.

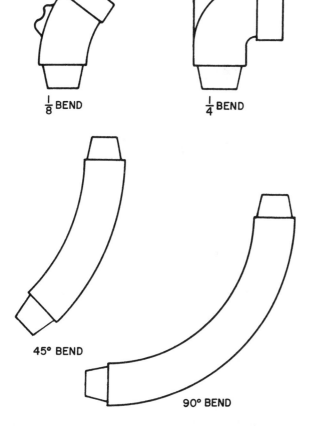

$\frac{1}{8}$ BEND $\qquad$ $\frac{1}{4}$ BEND

45° BEND

90° BEND

Fig. 2-2 Orangeburg fittings

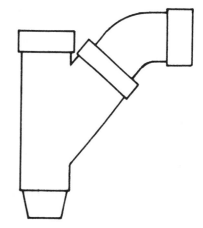

Fig. 2-3 Combination wye and 1/8 bend

REVIEW QUESTIONS

Multiple Choice

Select the best answer for each question.

1. Bituminized-fiber pipe is made from
   a. wood fibers and tar.
   b. tar-coated clay.
   c. ABS plastic.
   d. vitrified clay.

2. Which of the following bituminized-fiber pipe sizes and lengths is not correct?

    a. 2" diameter by 5' long

    b. 3" diameter by 5' long

    c. 4" diameter by 8' long

    d. 5" diameter by 5' long

3. A length of 4-inch bituminized-fiber pipe weighs how many pounds?

    a. 6.6 pounds

    b. 11.6 pounds

    c. 14.6 pounds

    d. 21.6 pounds

4. How are fiber pipe and fittings joined together?

    a. A standard plastic pipe coupling is used.

    b. A coupling with a 2-degree internal taper is used.

    c. Portland cement and straight-wall couplings are used.

    d. Threaded couplings are used.

5. How are small changes in direction made in the pipe trench?

    a. Shoring jacks are used to bend the pipeline.

    b. The pipe may be pried over with a shovel and then held in place with rocks.

    c. 5-degree couplings are used to make small changes in direction.

    d. The pipe is bent with heat.

6. Another name for bituminized-fiber fittings is

    a. ABS fittings.

    b. Orangeburg fittings.

    c. tar fittings.

    d. terra-cotta fittings.

7. What is used to join bituminized-fiber pipe to other pipe material?

    a. Adapter

    b. Coupling

    c. Wye

    d. Stretcher

8. When a cut is made on the job, the pipe end must be

    a. carefully cleaned.

    b. taped.

    c. wrapped with oakum.

    d. field-machined.

9. Orangeburg pipe laid in a trench must be _____ by hand.

    a. painted

    b. bent

    c. backfilled

    d. none of the above

10. The purpose of the knobs on the fittings is to

    a. align the fitting when it was being molded.

    b. provide a safe place to strike the fitting when driving it on.

    c. provide a spot to place stone under when leveling.

    d. provide a handhold.

UNIT 3 ASBESTOS CEMENT PIPE

OBJECTIVES

After studying this unit, the student will be able to:

- describe the sizes and uses of asbestos cement pipe and fittings.

- make a joint on asbestos pipe.

ASBESTOS CEMENT PIPE

Asbestos cement pipe, sometimes called *transite pipe*, consists of asbestos fibers, cement, and silica. The ingredients are throughly mixed while dry, and a measured amount of water is added. It is then formed on a mandrel into different sizes of pipe. Finally, it is cured or hardened by air and then by steam. During the curing process, the silica combines with the free lime found in the cement.

Asbestos cement pipe does not corrode or rust. Its smooth inner surface permits ease of flow.

Asbestos cement pipe is made in 10-foot lengths. A special blade for cutting asbestos cement is used if available. Otherwise, it may be cut with a regular wood saw, but the ends of the pipe must be machined.

CAUTION: When handling, cutting, or machining any asbestos product, respirators and protective clothing must be worn.

ASBESTOS CEMENT COUPLINGS

Asbestos cement couplings consist of two rubber compression sealing rings. The couplings are machined to receive the rubber rings and to insure a tight joint, figure 3-1.

Asbestos cement pipe may be joined to cast iron, bituminized-fiber fittings, or vitrified clay pipe by using adapters.

Long runs of pipe may be bent to a certain extent at the coupling. Bends are made in 11 1/4, 22 1/2, and 45-degree angles. Tees, wyes, and several adapters are also made for sewer connections in 4, 5, and 6-inch sizes.

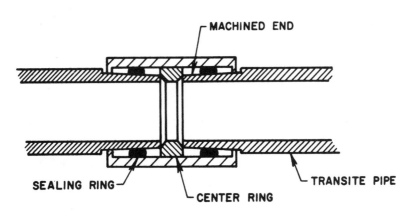

Fig. 3-1 Asbestos cement couplings

HOW TO MAKE A JOINT
ON ASBESTOS CEMENT PIPE

To make a joint on asbestos cement piping:

1. Place the sealing ring on the end of the pipe. Roll it back and forth to remove twists and square it with the pipe. Position it 1/4 inch from the end of the pipe.

2. Place the coupling against the sealing ring on the pipe and push it into place.

3. Place the sealing ring on the second length of pipe as in step 1.

4. Use an assembly tool to push the pipe into the coupling at the other end of the length of pipe, figure 3-2.

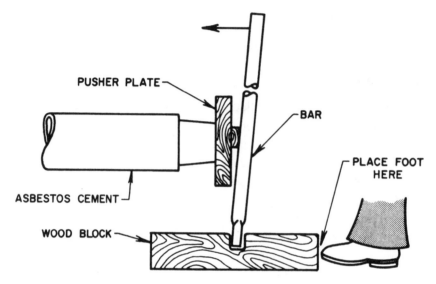

Fig. 3-2 Making a joint on asbestos cement pipe

REVIEW QUESTIONS

Multiple Choice

Select the best answer for each question.

1. Another name for asbestos cement pipe is
   a. Orangeburg pipe.
   b. transite pipe.
   c. cement pipe.
   d. PVC.

2. A length of asbestos cement pipe is
   a. 4 feet.
   b. 5 feet.
   c. 8 feet.
   d. 10 feet.

3. What must be worn when machining asbestos cement pipe?
   a. Loose clothing
   b. Respirator
   c. Apron
   d. None of the above

4. Asbestos cement pipe may be joined to other pipe material by using
   a. reducers.
   b. adapters.
   c. clamps.
   d. sleeves.

5. Lengths of pipe are joined together with couplings and
   a. pipe cement.
   b. pipe adapters.
   c. rubber rings.
   d. bearings.

6. Long runs of pipe may be bent slightly at the
   a. seams.
   b. wyes.
   c. tees.
   d. couplings.

7. Bends are available in angles of 22 1/2 degrees, 45 degrees, and
   a. 5 degrees.
   b. 7 1/2 degrees.
   c. 11 1/4 degrees.
   d. 92 degrees.

8. How far is the sealing ring positioned from the end before assembly?
   a. 4 inches
   b. 2 1/8 inches
   c. 3/4 inch
   d. 1/4 inch

9. The purpose of rolling the rubber ring back and forth on the pipe end is to
   a. remove twists.
   b. spread the joint compound.
   c. remove imperfections.
   d. stretch the ring.

10. After a piece of pipe is cut on the job, what must be done?
    a. It must be field-machined.
    b. The pipe must be cleaned.
    c. Nothing, it is ready to be assembled.
    d. It must be painted.

UNIT 4 GLASS PIPE AND FITTINGS

OBJECTIVES

After studying this unit, the student will be able to:

- discuss the features, sizes, and uses of glass pipe.

- install glass fittings.

GLASS PIPE

Glass pipe has been used for over 30 years. Until recent improvements, however, it was mainly used in chemical laboratories. Today glass pipe, such as Pyrex®, is used in a wide variety of industries: chemical, pharmaceutical, food, dairy, plating, pulp and paper, and textile.

One of the major advantages of Pyrex pipe is its resistance to corrosive acids and acidic materials. Its transparency allows colored liquids to be seen. This makes it easier to recover valuable materials and locate stoppages. Glass pipe insures product purity. The pipe does not corrode so it does not contaminate the products it carries. Another advantage is its smooth surface. Friction loss is only 80 to 90 percent that of clean steel pipe.

Pyrex pipe is tough and is highly heat-resistant. It can withstand internal pressure of 50 pounds per square inch and resist heat to 450 degrees Fahrenheit. Even higher pressures and temperatures are possible using special gaskets. Figure 4-1 gives the stock lengths and pipe sizes available.

JOINING GLASS PIPE AND FITTINGS

Pyrex pipe and fittings are joined with a coupling which requires no special treatment

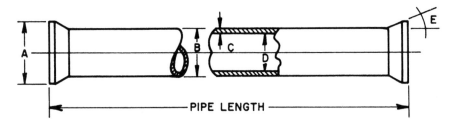

D. Pipe Size (Nominal I.D.)	A	B (O.D.)	C (Wall)	E (Angle)	Minimum Length	Stock Lengths	Approx. Weight Lbs. Per Foot
1''	1 9/16''	1.313''	0.156''	12°	3''		0.6
1 1/2''	2 1/8''	1.844''	0.172''	12°	3 1/2''	6'', 12'', 18'', 24'', 30'', 36'', 48'', 60'', 72'', 84'', 96'', 108'', 120'', in all diameters. Also 6'' by 8'' stocked (Priced same as 6'' by 12'').	1.0
2''	2 5/8''	2.344''	0.172''	12°	4''		1.13
3''	3 25/32''	3.406''	0.203''	12°	5''		2.0
4''	5 11/32''	4.530''	0.265''	21°	6''		3.4
6''	7 17/32''	6.656''	0.328''	21°	6''		6.3

Fig. 4-1 Glass pipe lengths and sizes

Fig. 4-2 Joining Pyrex pipe and fittings

on the cut end. The clamp is placed on the pipe and drawn tight with pliers, figure 4-2. Beveled spacers, figure 4-3, allow pitch and drainage where necessary. With a combination of lengths and spacers, any linear measurement can be obtained. By accurately estimating the piping for a job, cutting pipe and shaping

the ends on the job may not be necessary.

Pyrex fittings are very similar to metal pipe fittings, but there are not as many available. The manufacturer's catalog lists the fittings available and their sizes.

Pyrex pipe and fittings are joined with flared conical ends to take a special metal

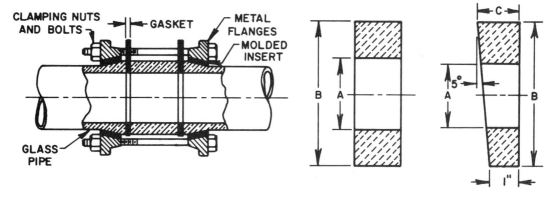

STANDARD SPACER

Pipe Size, A	B	Stock Lengths
1"	1 9/16"	1/2", 3/4", 1", 1 1/2", 2", 2 1/2", 3", 4", 5"
1 1/2"	2 5/32"	1/2", 3/4", 1", 1 1/2", 2", 2 1/2", 3", 4", 5"
2"	2 21/32"	1/2", 3/4", 1", 1 1/2", 2", 2 1/2", 3", 4", 5"
3"	3 27/32"	1/2", 3/4", 1", 1 1/2", 2", 2 1/2", 3", 4", 5"
4"	5 11/32"	1/2", 3/4", 1", 1 1/2", 2", 2 1/2", 3", 4", 5"
6"	7 1/2"	1/2", 3/4", 1", 1 1/2", 2"

BEVELED SPACER

Size A	B	C
1"	1 37/64"	1 9/64"
1 1/2"	2 5/32"	1 3/16"
2"	2 43/64"	1 15/64"
3"	3 27/32"	1 21/64"
4"	5 21/64"	1 15/32"

Fig. 4-3 Beveled spacers

Fig. 4-4 Exploded view of assembly

flange, figure 4-4. The flange is cushioned from the glass by molded asbestos inserts. Tightening the bolts grips the gasket between grooved pipe ends. There is no contact between the metal flange and the fluid in the pipe. The manufacturer's catalog gives the complete installation procedure for assembling the pipe and fittings.

Special gaskets are available in neoprene, gum rubber, asbestos, Koroseal®, and Teflon. Figure 4-5 suggests certain uses. The manufacturer will recommend uses in unusual conditions.

When glass pipe is connected to other materials, such as metal or copper tubing, a variety of adapters may be used. Figures 4-6 through 4-9 show some connections and adapters which must be installed.

GASKET THICKNESSES

Pipe Size	Asbestos, Neoprene, Koroseal, Gum Rubber	Teflon Type T	
1″	1/8″	1/16″	
1 1/2″	1/8″	1/16″	Use Style 1 gaskets to fit Style 1 flanges.
2″	1/8″	1/16″	Style 2 gaskets fit Style 2 flanges.
3″	1/8″	5/64″	Style 1-2 gaskets fit either style flange.
4″	1/8″	1/8″	
6″	1/8″	3/16″	

Type	Service	Maximum Woring Temp.
Asbestos Type A	Organic acids, solvents, chlorinated compounds and mineral acids	400° F.
Gum Rubber Type R-1	Food products, mineral acids except HNO_3	200° F.
"Neoprene" Type R-2	Mineral acids except HNO_3, distilled water	220° F.
Goodrich "Koroseal" Type R-3	Mineral acids including HNO_3, chlorinated compounds	150° F.
"Teflon" Type T	All	450° F.

Fig. 4-5 Gasket uses

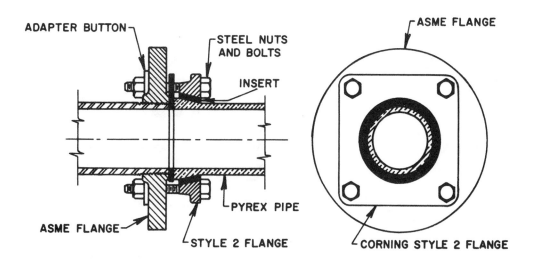

Fig. 4-6 Pyrex pipe using ASME flange

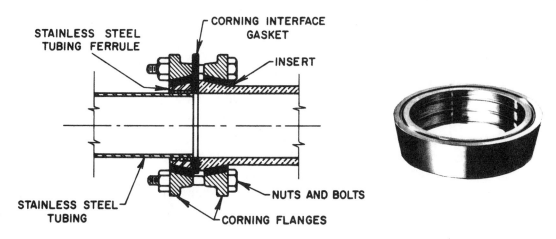

Fig. 4-7 Pyrex pipe connected to metal tubing

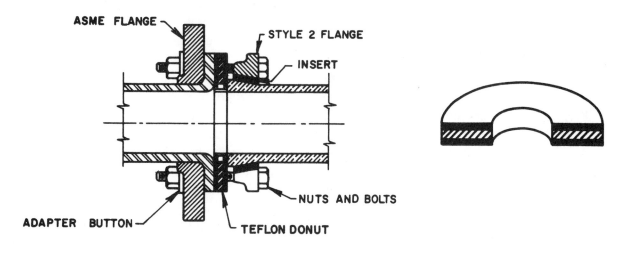

Fig. 4-8 Pyrex pipe connected to glass-lined steel

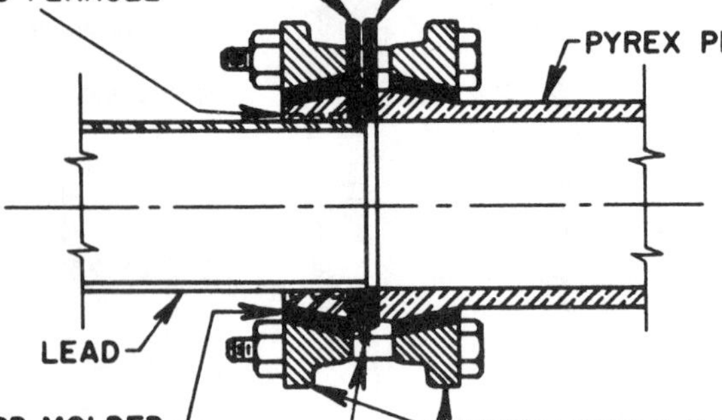

Fig. 4-9 Pyrex pipe connected to unthreaded lead pipe.

Labels in figure:
STAINLESS STEEL OR BRASS FERRULE
NEOPRENE OR GASKET (DOES NOT CONTACT FLUIDS)
GASKET TO CONTACT FLUIDS
PYREX PIPE
LEAD
STANDARD MOLDED INSERT
PYREX STYLE 2 FLANGE
FLARE OUT LEAD HERE

INSTALLING GLASS PIPE

Installing Pyrex pipe is similar to installing metal pipe in many ways. There are several exceptions, however, which must be considered to obtain a permanent glass pipe job.

While Pyrex pipe is tough, it must be handled carefully and installed without strain on the pipe. Glass pipe and fittings should be kept in their boxes or crates until used. If the pipe or fittings are scratched or nicked, the glass is weakened at that point and may break later.

Pyrex pipe can be used underground if it is insulated with material recommended by the manufacturer. It may be concealed if removable panels are used. If laid on floors, near doors, or in other exposed places, it should be protected from damage by using wire mesh covering, angle irons, or channels. Pyrex pipe systems must also be protected from damage against pressure surges. This is done by installing relief valves or, with piston pumps, by installing air chambers.

Valves, strainers, meters, and equipment must be rigidly supported apart from the pipe. Flexible elbows, Teflon bellows, or rubber hose are connected to equipment, particularly where vibration occurs. Valves are located on all branches, tanks, and pieces of equipment.

There is only one anchor point in a line of glass pipe unless an expansion loop or swing joint is provided. Anchors or hangers are lined with sheet rubber or asbestos to prevent scratching the pipe. Hangers are placed about 12 inches from each end on a 10-foot length of pipe. The best hangers permit the pipe to move 1 inch lengthwise or sidewise, figure 4-10.

Expansion of Pyrex pipe is approximately one-fourth that of steel pipe. It expands 0.000 001 8 inch per degree per inch. This is approximately 1/4 inch per 100 feet of pipe per 100-degree temperature difference.

Horizontal branches should not be supported closer than 8 feet to a riser to permit the riser to expand. Vertical risers or stacks are supported at the base using a padded saddle support or, in some places, by beam clamps under a flange.

To hang a line of glass pipe, start at a fixed point and hang, align, grade, and tighten one length of pipe at a time. Clevis hangers are excellent for use as they are flexible and can be adjusted for grade.

A special kit is available to prepare Pyrex pipe on the job so that the pipe may be cut. No further treatment is necessary.

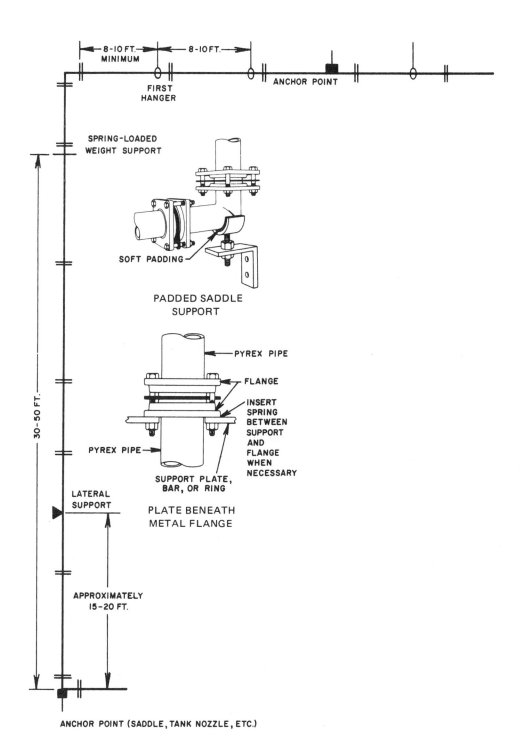

Fig. 4-10 Supporting glass/pipe

Pyrex piping is tested for stress by shaking it or by opening a flanged joint. To be certain that joints are tight, they are tested by using water pressure 1 1/2 times greater than the anticipated working pressure. An air test is not recommended.

17

SAFETY PRECAUTIONS

Pyrex pipe is often used to carry sulfuric acid, nitric acid, and other dangerous chemicals. Plumbers working on such lines must take every precaution to protect themselves and others from being burned. Acid burns are extremely painful and can disfigure the body. *If acid contacts the skin, immediately wash the area with plenty of water.*

Before working on pipe which has contained acid, be absolutely sure the valve is tightly closed. Next, flush out the pipe with water. *Wear rubber gloves, a mask, and goggles when flushing pipes.* Only when the pipe has been thoroughly flushed out is it safe to dismantle the line.

REVIEW QUESTIONS

Multiple Choice

Select the best answer for each question.

1. Which of the following is not an advantage of glass pipe?
   a. Corrosive resistance
   b. Product purity
   c. Strength
   d. Locating stoppages

2. Glass pipe may withstand an internal pressure of
   a. 50 psi.
   b. 75 psi.
   c. 100 psi.
   d. 125 psi.

3. Glass pipe will withstand heat up to
   a. 212 degrees.
   b. 250 degrees.
   c. 350 degrees.
   d. 450 degrees.

4. Pitch and drainage may be attained with _____ if necessary.
   a. beveled spacers
   b. drainage fittings
   c. pipe bending
   d. none of the above

5. Which of the following is not available as gasket material for glass pipe?
   a. Asbestos
   b. Neoprene
   c. Teflon
   d. Pyrex glass

6. Both _____ and _____ may cause glass pipe to break after installation.
   a. acids, oil
   b. nicks, scratches
   c. valves, strainers
   d. none of the above

7. Which of the following testing procedures is not recommended for glass pipe installations?
   a. Air pressure
   b. Water pressure
   c. Shaking
   d. Visual inspection

8. The expansion of glass pipe is approximately
   a. 4 times that of steel pipe.
   b. 1/2 that of steel pipe.
   c. 1/4 that of steel pipe.
   d. the same as steel pipe.

9. Hangers for glass pipe must be
   a. cast iron.   c. very tight.
   b. padded.   d. rigid.

10. When working on a pipeline which carries acid,
    a. no precautions need to be taken.
    b. loosen all hangers first.
    c. flush the system completely with water.
    d. none of the above

11. If the skin comes into contact with acid, the first thing to do is to
    a. flush the area with water immediately.
    b. go to a doctor immediately.
    c. bandage the area.
    d. report the accident to the supervisor.

12. What is the end-to-center measurement of a 6-foot length of pipe and a 6-inch, 90-degree Pyrex elbow?
    a. 6'-6"
    b. 6'-9"
    c. 6'-9 3/16"
    d. There is not enough information given to answer the question.

13. Hanger spacing for glass pipe should be
    a. 8 to 10 feet.   c. every 20 feet.
    b. 4 to 6 feet.   d. 18 to 24 inches.

14. If glass pipe is laid upon or near the floor, it should be
    a. tinted.
    b. protected with wire mesh or steel angle.
    c. insulated.
    d. heated first.

15. The weight of 2-inch glass pipe is
    a. 0.172 pound per foot.   c. 3.11 pounds per foot.
    b. 1.13 pounds per foot.   d. 11.5 pounds per foot.

UNIT 5 WELDED PIPING

OBJECTIVES

After studying this unit, the student will be able to:
- describe welding processes and types of welds.
- prepare pipe ends for welded fittings.
- properly align pipes for welding.

WELDING PROCESSES

There are two welding processes for welded fittings: gas welding and electric arc welding.

Gas Welding

Gas welding mixes acetylene and oxygen to create temperatures high enough to melt steel. The gas welding setup is similar to the soldering operation except there are two tanks and two hoses leading to the torch tip. The pure oxygen supplied by one tank permits much higher temperatures than if acetylene gas alone were used.

In operation, the flame melts the base metal together; that is, it melts the pipe metal to the fitting metal. Some filler metal is added with a steel welding rod much the same as soldering. The gas welding apparatus may be fitted with a cutting attachment. The cutting torch is used to raise the temperature until the pipe melts. Then a stream of pure oxygen is directed into the molten puddle. The steel actually burns. If the operator moves the torch around the pipe circumference, it cuts through the pipe.

Electric Arc Welding

Arc welding uses electric current and is less expensive than gas welding. The electricity for arc welding may be either direct current or alternating current. If it is direct current, it may have a positive or negative ground.

The type of welding electrode used for filler metal must be matched to the machine, the type of electricity used, and the base metal to be welded. Manufacturers' booklets show the proper rods and machine settings to use for a variety of conditions.

TYPES OF WELDS

A properly welded joint is leakproof, vibration proof, and stronger than the pipe itself. Many underground pipelines have welded joints. Hydraulic and steam lines are welded with increasing frequency. Almost any kind of metal can be joined by welding.

There are two basic types of weld forms: the butt weld and the fillet weld, figure 5-1. The *butt weld* is used to join metal that is laid edge to edge. The *fillet weld* is used to join metal with faces which form an angle.

Fittings are either fillet-welded or butt-welded. A fillet weld-type fitting is similar to a copper sweat fitting. The sweat fitting is soldered so that the solder flows to the bottom of the socket. The fillet weld fitting is joined only at the face of the fitting and to the pipe.

More and more copper-to-brass fittings are being brazed. In *brazing,* the joint is welded with a brass rod rather than soldered. Brass flows like solder to the depth of the socket. While it is possible to weld cast iron and malleable iron by welding with iron rods, brazing these materials seems to give a more satisfactory job.

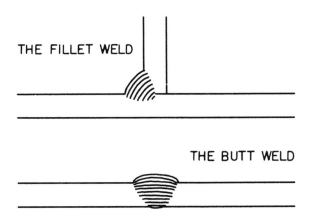

Fig. 5-1 Types of welds

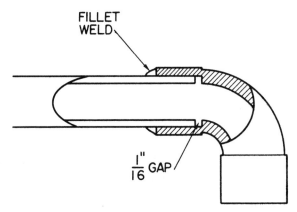

Fig. 5-2 Fillet weld on socket fitting

Fillet Weld End Preparation

A fillet weld is used with a socket fitting. A socket fitting is self-aligning. The end of the pipe is cut square and reamed. The pipe is inserted into the fitting and marked at the face of the fitting. The pipe is then withdrawn 1/16 inch from the fitting. The 1/16 inch allows the heated pipe to expand into the fitting without disturbing the joint as it is being welded, figure 5-2. The fitting allowance for a socket type, 90-degree fitting equals the pipe size.

Butt Weld End Preparation

The butt weld fitting has the same inside and outside diameter as the pipe to which it is joined, figure 5-3. Long-radius and short-radius patterns are available for 90-degree elbows. This is similar to the long sweep and the short sweep encountered in soil fittings.

The allowance for the short-radius elbow is equal to the pipe size. The long-radius allowance is equal to 1 1/2 times the pipe size. The allowance for a 45-degree elbow is equal to 5/8 times the pipe size. An additional allowance must be made at the joint to allow the weld to penetrate to the inside diameter of the pipe. A 1/16-inch separation is used for standard-weight pipe.

Before the joint is set up for tacking, the pipe is beveled to the same angle as the fitting, see figure 5-3. A *tack* is a short weld

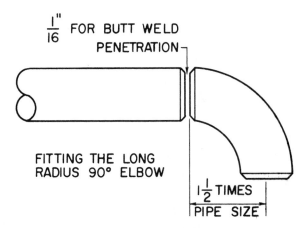

Fig. 5-3 Butt weld

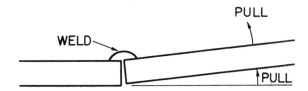

Fig. 5-4 When the weld cools, the metal attempts to pull toward the welded side.

which holds the fitting in line while the weld is being made. When tacking fittings to pipe, remember that the joint will pull or shrink towards the weld as the weld cools, figure 5-4.

21

PIPE ALIGNMENT

Proper alignment of piping is very important when welding. If done correctly, welding is easier to do and the piping system will function properly.

There are many ways to align piping. Most are done using framing squares, levels, and rules. The following procedures suggest methods of obtaining a good alignment quickly.

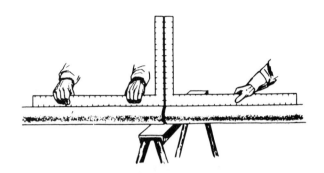

Fig. 5-5 Pipe-to-pipe alignment

Pipe-to-Pipe Alignment

Move the pipe lengths together until the bevels nearly abut or touch, allowing a gap for the weld (1/16 inch), figure 5-5. Center squares on top of both pipes and move the pipe up and down until the squares are aligned. When the pipes are aligned, the numbers on the two squares will match up. Tack weld the top and bottom. Repeat the procedure by placing the squares on the side of the pipe. Correct the alignment by moving the pipe left or right. Tack weld each side.

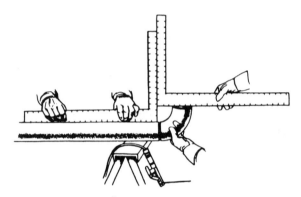

Fig. 5-6 90° elbow-to-pipe alignment

90-Degree Elbow-to-Pipe Alignment

Place the fitting's bevel so it nearly touches the bevel of the pipe, figure 5-6. Allow a gap for welding. Tack weld on top. Center one square on top of the pipe. Center the second square on the elbow's alternate   face. Move the elbow until the squares are aligned.

45-Degree Elbow-to-Pipe Alignment

The procedure for aligning a 45-degree elbow to pipe is the same as aligning a 90-degree elbow to pipe except the squares are crossed, figure 5-7. To obtain a correct 45-degree angle, move the squares until the same numbers on the inside scales are aligned.

Alternate method. Abut the fitting and the pipe allowing a gap for the weld. Center a spirit level on the pipe, figure 5-8. Center a 45-degree spirit level on the face of the elbow. Move the elbow until the 45-degree bubble is centered.

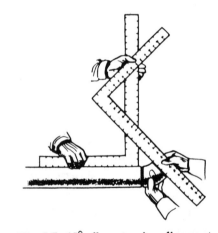

Fig. 5-7 45° elbow-to-pipe alignment

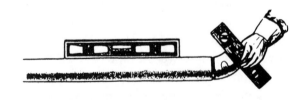

Fig. 5-8 45° elbow-to-pipe alternate method

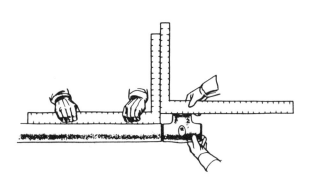

Fig. 5-9 Tee-to-pipe alignment

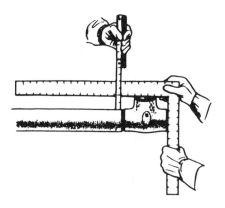

Fig. 5-10 Tee-to-pipe alternate method

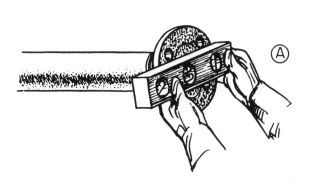

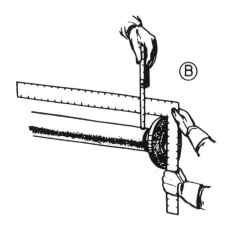

Fig. 5-11 Flange-to-pipe alignment

Tee-to-Pipe Alignment

Abut the bevels, allowing a gap for the weld. Tack weld on top. Center a square on top of the pipe, figure 5-9. Place the second square on the center of the branch outlet. Move the tee until the squares are aligned.

Alternate method. Abut the bevels. Place a square on the tee, figure 5-10. Center a rule on top of the pipe. The blade of the square should be parallel with the pipe. Check this by measuring with the rule at several points along the pipe.

Flange-to-Pipe Alignment

Abut the flange to the pipe. Align the top two holes of the flange with a spirit level, figure 5-11(A). Move the flange until the bubble is centered. Tack weld on top. Center the square on the face of the flange, figure 5-11(B). Center the rule on top of the pipe. Move the flange until the square and pipe are parallel. Tack weld on bottom. Center the square on the face of the flange. Center a rule on the side of the pipe and align as before. Tack both sides.

REVIEW QUESTIONS

Multiple Choice

Select the best answer for each question.

1. Gas welding uses _____ and _____ to obtain high temperature
   a. acetylene, air
   b. natural gas, hydrogen
   c. oxygen, acetylene
   d. methane, oxygen

2. The metal of the pipe and the fitting is called
   a. base metal.
   b. weld reinforcement.
   c. root bead.
   d. none of the above.

3. Steel pipe may be _____ welded or _____ welded.
   a. arc, gas
   b. hammer, gas
   c. arc, chemical
   d. chemical, gas

4. The two types of welds are
   a. root and butt.
   b. reinforcement and pass.
   c. fillet and butt.
   d. upbead and forehand.

5. A socket fitting requires a
   a. fillet weld.
   b. solder weld.
   c. butt weld.
   d. backhand weld.

6. The fitting allowance for a 2-inch, 90-degree elbow for a fillet weld is
   a. 1/2 inch.
   b. 1 1/8 inches.
   c. 1 1/2 inches.
   d. 2 inches.

7. The fitting allowance for a 3-inch, 90-degree long radius elbow for a butt weld is
   a. 1 inch.
   b. 1 1/2 inches
   c. 3 inches.
   d. 4 1/2 inches.

8. Framing squares may be used to
   a. support the pipe.
   b. align the pipe and fittings.
   c. hang the pipe.
   d. none of the above.

9. A short weld that holds the fitting in line is called a
   a. root bead.
   b. tack weld.
   c. reinforcement.
   d. solder weld.

10. The gap between butt weld fittings and pipe is about
    a. 1/16 inch.
    b. 1/4 inch.
    c. 2 inches.
    d. 2 3/4 inches.

SECTION **2**

Soil Pipe

UNIT 6 DRAINAGE TRAPS

OBJECTIVES

After studying this unit, the student will be able to:

- discuss the uses of drainage traps.
- identify the sizes and types of drainage traps.

DRAINAGE TRAPS

Cast-iron galvanized drainage traps are used in large buildings where pipes are exposed on ceilings and below fixtures. They are used most often on urinals and floor drains in public rest rooms. However, they may also be used on any drainpipe above ground or inside buildings.

Drainage traps range in sizes from 1 1/4 to 10 inches and come in several patterns. Some have cleanouts and others have vent connections. Figure 6-1 shows four types of drainage traps.

Carefully check the state and local codes concerning the use of these traps as some of them may be specifically forbidden. Full S traps and 3/4 S traps, popular when the crown method of venting was permitted, are easily siphoned.

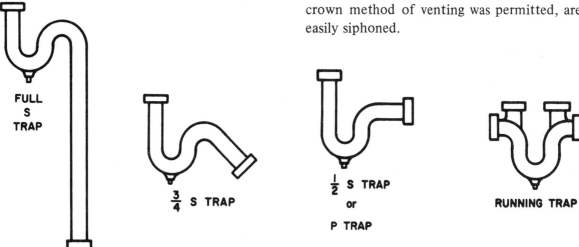

Fig. 6-1 Drainage traps

REVIEW QUESTIONS

Multiple Choice

Select the best answer for each question.

1. A plug in the bottom of a drainage trap
   a. provides an outlet for a future connection.
   b. provides an outlet to clean out the trap.
   c. anchors the trap in the mold while it is being cast.
   d. none of the above.

2. Cast-iron galvanized traps are found mostly in
   a. more expensive homes.
   b. modular homes.
   c. cesspools.
   d. large commercial buildings.

3. What kind of pipe is used with cast-iron galvanized drainage traps?
   a. Galvanized steel
   b. Copper DWV
   c. Chrome-plated brass
   d. Plastic

4. What can be a problem with S and 3/4 S traps?
   a. Lack of durability
   b. Loss of trap seal due to back pressure
   c. Loss of trap seal due to siphonage
   d. Obtaining materials

5. What is the smallest size the cast-iron galvanized trap is made in?
   a. 1 1/4 inches
   b. 1 1/2 inches
   c. 2 inches
   d. 3 inches

6. What type of trap is used to drain a sink to the floor?
   a. S trap
   b. 3/4 S trap
   c. 1/2 S trap
   d. Running trap

7. What type of trap is used in a horizontal line?
   a. S trap
   b. 3/4 S trap
   c. 1/2 S trap
   d. Running trap

8. Which trap is used to drain a sink to the wall?
   a. S trap
   b. 3/4 S trap
   c. 1/2 S trap
   d. Running trap

9. What is another name for the 1/2 S trap?
   a. Running trap
   b. Deep-seal trap
   c. Semi-S trap
   d. P trap

10. The end preparation for drainage traps is
    a. sweated.
    b. threaded.
    c. welded.
    d. none of the above.

UNIT 7 FLOOR AND AREA DRAINS

OBJECTIVES

After studying this unit, the student will be able to:

- discuss the types and uses of floor and area drains.
- calculate the size of area drains.

FLOOR AND AREA DRAINS

Cast-iron floor and area drains are available in a wide variety of styles and patterns. Floor drains carry away any leakage or spilled water in such places as elevator pits, garages, boiler rooms, and laundries. Area drains are placed in driveways and other paved surfaces. The top is removable for cleaning purposes.

Floor and area drains installed in driveways or garages are placed flush with the pavement or floors and are extra heavy to withstand traffic, figure 7-1. Floor drains with flanges are required in some areas of the United States, figure 7-2.

All floor and area drains must have a trap on the waste pipe that can be cleaned in case of stoppage, figure 7-3. The trap must be below the frost line to prevent freezing. If the trap is directly under the drain, it may be cleaned by removing the grate. If the distance is not too great, the area drain trap is placed in the basement of the building. This prevents freezing or placing it below the frost line. Those floor drains which are caulked to the soil pipe are best.

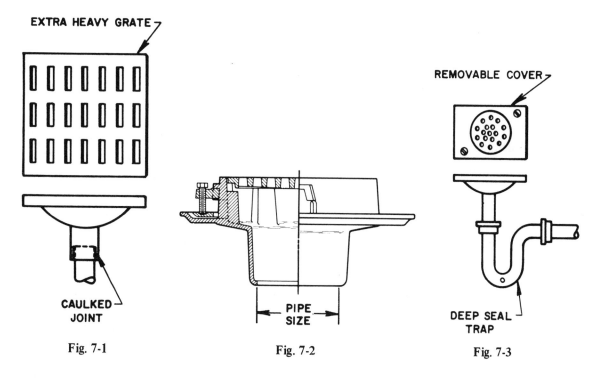

| Fig. 7-1 | Fig. 7-2 | Fig. 7-3 |

Area drains have a deep seal trap to prevent the seal from evaporating in periods of drought. Floor drains must have a water supply handy in order to replace the water in the seal which evaporates from the heat in enclosed areas.

SIZING AREA DRAINS

The size of area drains for large surfaces may be determined by using the table in figure 7-4, Size of Horizontal Storm Drains. A 2-inch drain is the minimum size that may be used. This is sufficient for areas up to 100 square feet.

The table in figure 7-4 is designed for rainfall rates of 4 inches per hour. To convert to your area, multiply the answer by the number of inches per hour for your area.

SIZE OF HORIZONTAL STORM DRAINS

Diameter of Drain Inches	Maximum Projected Area for Drains of Various Slopes					
	1/8 in. Slope		1/4 in. Slope		1/2 in. slope	
	Square Feet	gpm	Square Feet	gpm	Square Feet	gpm
3	822	34	1160	48	1644	68
4	1880	78	2650	110	3760	156
5	3340	139	4720	196	6680	278
6	5350	222	7550	314	10700	445
8	11500	478	16300	677	23000	956
10	20700	860	29200	1214	41400	1721
12	33300	1384	47000	1953	66600	2768
15	59500	2473	84000	3491	119000	4946

Based on rainfall rate of 4" per hour

Fig. 7-4

REVIEW QUESTIONS

Multiple Choice

Select the best answer for each question.

1. The purpose of a floor drain is to
   a. carry away spilled water and leakages.
   b. carry the discharge from washing machines.
   c. carry away storm water.
   d. support the main house drain.

2. What type of floor drain is used in a driveway?
   a. One with a removable cover
   b. One with an antisiphon grate
   c. The antibackflow type
   d. None of the above

3. What is the advantage of placing the trap directly below the drain?

   a. A savings of material   c. Better access for cleaning

   b. Greater support for the grate   d. Less chance of freezing

4. Area drains used in driveways have a

   a. hose bib.   c. integral trap.

   b. cement cover.   d. heavy grate.

5. Area drains have a

   a. deep-seal trap.   c. double-cleanout trap.

   b. cast-bronze trap.   d. drum-type trap.

6. Traps placed on area drains must be protected from

   a. oxidation.   c. freezing.

   b. crushing.   d. acids.

7. The minimum size pipe to an area drain is

   a. 1 inch.   c. 1 1/2 inches.

   b. 1 1/4 inches.   d. 2 inches.

8. What size pipe is used to drain an area 10 feet wide by 100 feet long? Rainfall is 4 inches per hour.

   a. 2-inch pipe at 1/2-inch slope   c. 3-inch pipe at 1/8-inch slope

   b. 3-inch pipe at 1/4-inch slope   d. 4-inch pipe at 1/4-inch slope

9. With a maximum slope of 1/4 inch per foot, what is the minimum size pipe to use to drain 6700 square feet?

   a. 3-inch pipe   c. 5-inch pipe

   b. 4-inch pipe   d. 6-inch pipe

10. If there are to be two drains at a 1/4 inch per foot slope in a small parking area (20′ x 100′), what size should they be?

   a. 1 1/2 inches   c. 3 inches

   b. 2 inches   d. 5 inches

UNIT 8 ROOF FLASHINGS

OBJECTIVES

After studying this unit, the student will be able to:

- discuss the purpose of roof flashings and where they may be used.
- describe how to make a flashing.

ROOF FLASHINGS

Vent pipes must be watertight where they pass through roofs. This is done with a *roof flashing*, also called a *roof flange*.

In most cases an adjustable roof flange is used on the job, figure 8-1. These may be adjusted to a wide variety of angles, from a flat roof to one with considerable pitch. The adjustment in pitch angle is made by rotating the body of the flange. Some have a neoprene ring where the pipe passes through. Some have a lead ring. The lead ring is tapped tightly against the pipe.

Sheet lead or copper is often used to make a flashing on the job. A handmade flashing of sheet lead is shown in figure 8-2. Notice that the flashing is caulked right into the lead joint. Care must be taken to avoid cutting through the sheet lead with the irons.

Figure 8-3 shows the layout patterns for the cone-shaped part of a handmade flashing. These are then soldered to a piece of flat lead to go against the roof.

On pitched roofs, the upper part of the flashing is slipped under the roofing material. The lower part is placed over the roofing material. A flexible roofing cement is then applied to the seal, to the pipe, to the nails, and all around the edge of the flange.

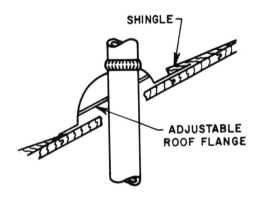

Fig. 8-1 Adjustable roof flange

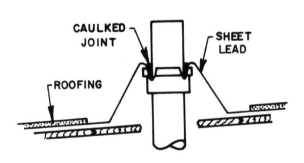

Fig. 8-2 Sheet lead flashing

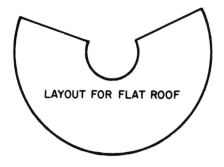

Fig. 8-3 Flashing layout patterns

REVIEW QUESTIONS

Multiple Choice

Select the best answer for each question.

1. What is the advantage of using an adjustable roof flange?

    a. It may be used for any size pipe.
    b. It may be adjusted for most roof slants.
    c. It is easy to make.
    d. It costs the least.

2. Why are roof flanges necessary?

    a. To keep water from leaking through the hole in the roof
    b. To support the top of the stack
    c. To provide for roof expansion
    d. To seal out sewer gas

3. What two materials are most common for handmade flashings?

    a. Galvanized sheet metal and rubber
    b. Wood and asbestos
    c. Roofing felt and aluminum
    d. Sheet lead and sheet copper

4. How is the flange installed on a sloped, shingled roof?

    a. Over the upper shingles and under the lower
    b. Under the upper shingles and over the lower
    c. Carefully nailed over the shingles
    d. Nailed beneath the roof

5. The layouts in figure 8-3 are for

    a. cutting the hole in the roof.
    b. cutting out the part that goes under the shingles.
    c. cutting out the part that goes around the pipe in a cone.
    d. none of the above.

6. After the flashing is installed, what must be done?

    a. The layout must be cut.
    b. The stack must be plugged.
    c. The rest of the stack must be installed.
    d. The flange must be sealed with roofing cement.

7. How is the handmade flange in figure 8-2 made watertight?

    a. By caulking the sheet lead into the lead joint
    b. By tapping the lead against the pipe
    c. By gluing the roofing to the flange
    d. By soldering the flange to the stack

8. Another name for a roof flashing is

    a. a roof shingle.   c. a sheet lead layout.
    b. a roof flange.   d. a stack sealer.

UNIT 9 CLOSET BENDS

OBJECTIVE

After studying this unit, the student will be able to:

• describe the types and sizes of closet bends.

CLOSET BENDS

A closet bend connects the water closet to the drainage line. It is often made of cast iron. It has a 4-inch diameter and comes in many different lengths. A cast-iron flange is caulked onto the closet bend to secure the toilet bowl to the drainpipe.

The closet bend length can be adjusted by cutting at the grooves at either end. Closet bends also have side outlets or tappings on either side or both on the same side, figure 9-1.

Closet bends are specified by three dimensions. The first dimension gives the diameter. The second dimension gives the end-to-center measurement of the short end. The third dimension gives the end-to-center measurement of the long end. The closet bend in figure 9-2, page 34, is a 4"x6"x16" closet bend with 2 RH (right-hand) outlets.

Some localities prohibit cast-iron bends. They claim cast-iron bends are too rigid and may break the toilet bowl if the building or stack settles. Softer lead bends allow for this settlement without damage to the fixtures. However, lead bends often break at the joints and cause damage to the ceiling below.

Some areas also prohibit the use of closet bends with tappings, as in figure 9-1. Check with local codes before installation.

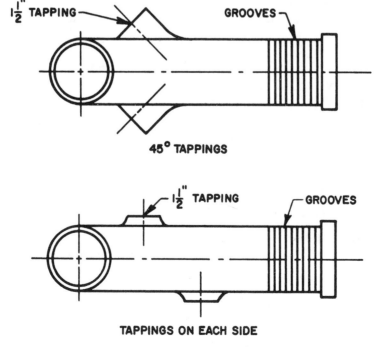

Fig. 9-1 Note tappings and grooves

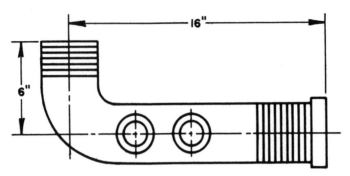

Fig. 9-2 Closet bend with right-hand tappings

REVIEW QUESTIONS

Multiple Choice

Select the best answer for each question.

1. In a closet bend measuring 4″ x 6″ x 18″, what does the 4″ represent?

   a. The diameter
   b. The length of the upturn
   c. The length of the long end
   d. The distance between outlets

2. In a closet bend measuring 4″ x 6″ x 18″, what does the 6″ represent?

   a. The diameter
   b. The length of the upturn
   c. The length of the long end
   d. The distance between outlets

3. In a closet bend measuring 4″ x 6″ x 18″, what does the 18″ represent?

   a. The diameter
   b. The length of the short end
   c. The length of the long end
   d. The distance between outlets

4. Specify a closet bend with one outlet on each side.

   a. A 4″ x 6″ x 16″ closet bend with 2 RH outlets
   b. A 4″ x 6″ x 16″ closet bend with R and L outlets
   c. A 4″ x 6″ x 16″ closet bend with 2 LH outlets
   d. None of the above

5. The grooves on the closet bend make it easier to

   a. align the bend.
   b. measure the bend.
   c. cut the bend.
   d. visualize the bend.

6. How is a toilet bowl attached to a cast-iron closet bend?

   a. By using a closet flange
   b. By gluing it with a wax ring
   c. By using a caulked lead joint
   d. By threading it directly

7. Why are cast-iron closet bends sometimes banned?

   a. They break at the outside of the bend.
   b. They tend to leak after a time.
   c. The grooves prevent a good caulked joint.
   d. The connection is too rigid.

8. What is a disadvantage of using a lead bend?

   a. It can crack and leak causing damage to the ceiling below.
   b. The connection between pipe and toilet is too rigid.
   c. There is too much weight on the ceiling joints.
   d. Electrolysis may occur.

9. What is an advantage of using a lead bend?

   a. It is easier to install than a cast-iron bend.
   b. It is not subject to freezing.
   c. It is softer and does not put strain on the toilet bowl.
   d. It may be stretched to fit.

10. How is the closet flange attached to the closet bend?

    a. With closet bolts   c. With a soldered joint
    b. With wood screws   d. With a caulked lead joint

UNIT 10 MISCELLANEOUS SOIL FITTINGS

OBJECTIVE

After studying this unit, the student will be able to:

- describe types of soil fittings.

SOIL PIPE FITTINGS

A *caulking sleeve*, figure 10-1, is used to join galvanized screw pipe to soil pipe. Often galvanized iron pipe is used on short branches to separate fixtures. One end of a caulking sleeve is similar to the bead end of a 2-inch length of soil pipe. The other end is tapped with a standard 1 1/4, 1 1/2, or 2-inch thread. The bead end has a lug which prevents it from being turned when the pipe is screwed on.

Vent cowls are put on the top of fresh air inlets when the vents are placed in a grass plot. They are made in 3, 4, 5, and 6-inch sizes in bead and bell patterns, figure 10-2. The use of vent cowls is prohibited in some areas.

A *fresh air inlet* with a dirt pocket is used when the fresh air inlet is located at the curb. It consists of a dirt pocket under the opening in the cover and is placed flush with the pavement. The lid is solid over the fresh air inlet so that dirt will not fall into the trap, figure 10-3.

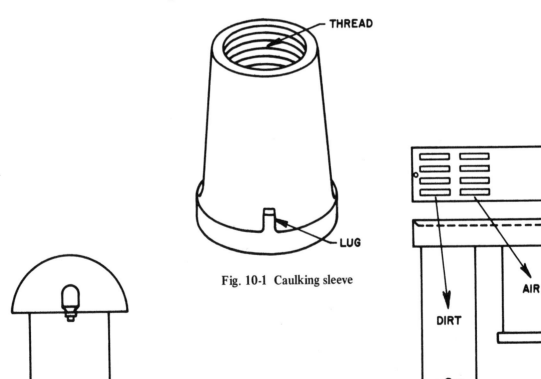

Fig. 10-1 Caulking sleeve

Fig. 10-2 Vent cowl

Fig. 10-3 Fresh air inlet

REVIEW QUESTIONS

Multiple Choice

Select the best answer for each question.

1. A 1 1/2-inch galvanized sink waste is to be connected to a 4" x 2" combination wye and 1/8 bend. Specify the fitting to use.

   a. A plastic transition fitting
   b. A 2" x 1 1/2" caulking sleeve
   c. A 2" x 1 1/2" reducing coupling
   d. A 2" x 1 1/2" union

2. A fresh air inlet is to be placed on a grass plot. Specify the fitting for the job.

   a. An adapter plug
   b. A curb box
   c. A flush-mount curb box
   d. A vent cowl

3. The purpose of the dirt pocket shown in figure 10-3 is to

   a. anchor the vent box firmly in the soil.
   b. strengthen the casting.
   c. prevent foreign objects from entering the drain.
   d. reduce the weight.

4. What is the purpose of the lug on the caulking sleeve?

   a. To provide a place to pick up the fitting
   b. To prevent cracking
   c. To give the pipe wrench a good grip
   d. To keep the fitting from being turned in the joint

5. The purpose of having slots in only one end of the cover in figure 10-3 is to

   a. allow light to pass through.
   b. prevent debris from entering the house sewer.
   c. indicate which way the drain is running.
   d. none of the above.

6. Where will the stoppage be if water is observed running out of the fresh air inlet?

   a. In the laundry tray
   b. In the vent itself
   c. Between the vent and the house
   d. Between the vent and the street

7. Where are fresh air inlets often located?

   a. Just inside the basement wall
   b. At the curb, flush with the pavement
   c. Below second-floor windows
   d. Above the roof

8. Which fitting may be prohibited in some areas?

   a. The vent cowl
   b. The caulking sleeve
   c. The flush-mount fresh air inlet
   d. None of the above

9. Which fitting joins soil pipe to a threaded joint?

   a. A vent cowl
   b. A threaded coupling

   c. A caulking sleeve
   d. A fresh air inlet

10. What does the term *tapped* mean?

   a. Pressed into the soil
   b. Cast with holes in it

   c. Internally threaded
   d. Tested for cracks

UNIT 11 BACKWATER VALVES

OBJECTIVE

After studying this unit, the student will be able to:

- describe the types and purpose of backwater valves.

BACKWATER VALVES

Backwater valves are check valves placed in drainpipes to prevent sewage from flowing backwards into buildings. Backwater valves must be installed where fixtures are placed in a basement, where sewage is pumped up to a house drain, or where buildings are near rivers.

The two types of backwater valves are the swing type and the balance type.

The *swing-type backwater valve*, figure 11-1, has a light, brass disc which is hinged at the top and closes against the brass seat. The disadvantage of the swing type is that it is closed except when water is flowing in the right direction. When it is closed for these long periods, the drainage system is not being ventilated.

The *balance-type backwater valve*, figure 11-2, has a disc on one end of an arm which is balanced in the center. An adjustable weight is placed on the opposite end to keep the valve open under normal use. Check with the local code before installing this valve.

In case of a stoppage in the drain or a flood, the water backing up closes the valve. Either valve is equipped with a large cleanout. A gate valve may be used on the drain as an emergency shutoff.

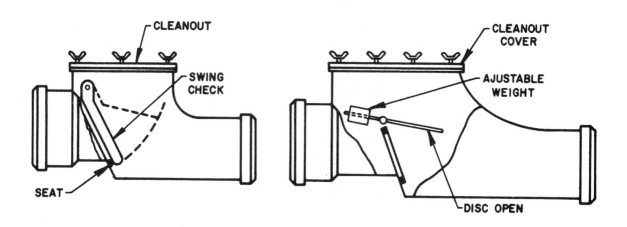

Fig. 11-1 Swing-type backwater valve   Fig. 11-2 Balance-type backwater valve

REVIEW QUESTIONS

Multiple Choice

Select the best answer for each question.

1. A laundry tray in a basement is protected with a
   - a. backwater valve.
   - b. metal cover.
   - c. relief valve.
   - d. coat of asphalt.

2. Another name for a backwater valve is a
   - a. relief valve.
   - b. gate valve.
   - c. bell-type tank siphon.
   - d. check valve.

3. One problem wth the swing-type backwater valve is that
   - a. the bearing surfaces wear out.
   - b. the moving parts are inaccessible.
   - c. it does not allow continuous ventilation of the system.
   - d. it may not be used in dry areas.

4. One advantage of the balance-type backwater valve is that
   - a. it allows continuous ventilation of the system.
   - b. it is simpler in design.
   - c. it presents no restriction at all to the flow.
   - d. it may be used in any position.

5. The purpose of a backwater valve is to
   - a. allow continuous ventilation of the system.
   - b. keep sewage from flowing backward into the building.
   - c. release sewage into the main sewer at predetermined times.
   - d. none of the above

6. If a backwater valve cover is not carefully replaced
   - a. raw sewage will leak out.
   - b. the gasket will be damaged.
   - c. the valve will not function.
   - d. the valve stem will corrode.

7. Fixtures in a basement are lower than the level of the building sewer and are pumped up to drain level. If there is no backwater valve and the sewage backs up, what will happen?
   - a. The fixtures will fill to the flood rim.
   - b. The sewage will run over the fixtures and into the cellar until the condition is corrected.
   - c. The water will come up in the fixtures until the pressure is equalized.
   - d. None of the above.

8. What actually causes a balance-type backwater valve to close?
   - a. Gravity
   - b. Unequal air pressures within the system
   - c. Back pressure within the system
   - d. Water backing up and pushing against the disc

9. What actually causes the swing-type backwater valve to close?

    a. Gravity

    b. Unequal air pressures within the system

    c. Back pressure within the system

    d. Water backing up and pushing against the disc

10. A backwater valve is required in

    a. a building in a valley.

    b. a building on a hill.

    c. a building served by an extremely deep sewer.

    d. a building along a river.

UNIT 12 INCREASERS AND REDUCERS

OBJECTIVE

After studying this unit, the student will be able to:

- describe how and where to increase or reduce drainpipes.

INCREASERS

Small-diameter vent pipes are likely to be closed by frost on the inside of the pipe in cold climates. This may cause loss of trap seals by siphonage and permit sewer gas to enter the building. To prevent this, the vent pipe is increased in size below the roof with pipe increasers, figure 12-1. The increaser in figure 12-2 is tapped for screw pipe on the small end.

A tapping boss, or projecting block, on 1 1/2″ x 2″ tapped increasers may be tapped for 1 1/4 to 2-inch pipe threads. Tappings are made according to American Standard Pipe Thread specifications.

REDUCERS

The most efficient horizontal pipe is not too large. Large drain lines may be reduced past a large branch on the upstream side.

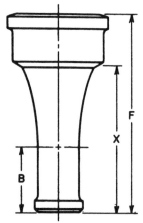

Increasers

Size	B	F	X[1]	Weight
2 x 3	4—	11 3/4	9	9
2 x 4	4—	12 –	9	10
2 x 5	4—	12 –	9	12
2 x 6	4—	12 –	9	13
3 x 4	4—	12 –	9	12
3 x 5	4—	12 –	9	14
3 x 6	4—	12 –	9	15
4 x 5	4—	12 –	9	15
4 x 6	4—	12 –	9	16
4 x 8	4—	15 1/2	12	32

Fig. 12-1 Increasers

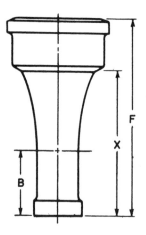

Tapped Increasers

Size	B	F	X[1]	Weight
1 1/2 x 2	4	10 1/2	8	7
2 x 3	4	11 3/4	9	9
2 x 4	4	12—	9	11
2 x 5	4	12—	9	12
2 x 6	4	12—	9	14

All dimensions given in inches. Weights given in pounds.

[1]Dimension X is the laying length.

Fig. 12-2 Tapped increasers

The run of soil fittings is never reduced. Therefore to reduce a main drainpipe, a reducer must be placed in the bell of the run of the branch fitting, figure 12-3. Reducers are made from 3 to 15 inches, figure 12-4.

A soil pipe increaser has the hub or bell on the larger end. A soil pipe reducer has a hub or bell on the smaller end.

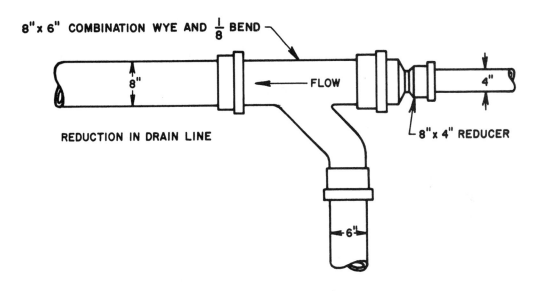

Fig. 12-3 Reduction in drain line

Reducers				
Size	B	F	X¹	Weight
3 x 2	3 3/4	7 1/4	4 3/4	6
4 x 2	4–	7 1/2	5–	7
4 x 3	4–	7 3/4	5–	9
5 x 2	4–	7 1/2	5–	8
5 x 3	4–	7 3/4	5–	10
5 x 4	4–	8–	5–	11
6 x 2	4–	7 1/2	5–	9
6 x 3	4–	7 3/4	5–	11
6 x 4	4–	8–	5–	12
6 x 5	4–	8–	5–	13

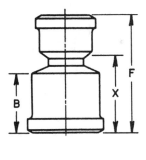

Fig. 12-4 Reducers

REVIEW QUESTIONS

Multiple Choice

Select the best answer for each question.

1. In running a 2-inch vent stack through a roof in some areas, the stack must be increased just under the roof to prevent freezing. What might happen if this precaution is not taken?
    a. Expansion and contraction of the vent may cause leaks around the roof flange.
    b. The cold air may not create a chimney effect and there would be no air currents in the stack.
    c. Frost may gradually collect in the stack and eventually close it up.
    d. None of the above.

2. Drainpipes may be reduced in size
    a. on the street side of the test tee.
    b. at branches and upstream of branches.
    c. downstream of the last fixture before exiting the building.
    d. none of the above.

3. If a 2-inch, galvanized steel vent stack is installed in Vermont, what fitting is used to increase it before it passes through the roof?
    a. A 3″ x 2″ reducer
    b. A 2″ x 3″ tapped increaser
    c. A 2″ x 3″ standard increaser
    d. An Orangeburg reducer

4. What is the actual laying length of a 2″ x 5″ increaser?
    a. 7 inches
    b. 9 inches
    c. 12 inches
    d. 15 inches

5. What happens when a vent pipe freezes close?
    a. Traps will be pulled out or siphoned.
    b. Pressure is created at the base of the stack.
    c. Water backs up into the lowest fixture.
    d. A pressure plenum forms above the slug of water.

6. The overall length of a 2″ x 4″ increaser is most nearly
    a. 10 inches.
    b. 12 inches.
    c. 15 inches.
    d. 25 inches.

7. The fitting allowance for a 2″ x 4″ increaser is
    a. 9 inches.
    b. 11 3/4 inches.
    c. 12 inches.
    d. 19 inches.

8. The weight of a 4″ x 5″ increaser is
    a. 5 pounds.
    b. 9 1/2 pounds.
    c. 12 pounds.
    d. 15 pounds.

9. The fitting allowance for a 5″ x 3″ reducer is
    a. 5 inches.   c. 11 inches.
    b. 10 inches.   d. None of the above.

10. The overall length of a 6″ x 5″ reducer is
    a. 7 1/2 inches.   c. 10 inches.
    b. 8 inches.   d. 13 inches.

UNIT 13 INSERTABLE JOINTS

OBJECTIVE

After studying this unit, the student will be able to:

- install an insertable fitting.

INSERTABLE JOINTS

In alteration or repair work, it is often necessary to insert a fitting or repair a broken length of existing soil line. This job may be difficult since there is no spring to a cast-iron pipe and the bells are 2 1/2 inches deep. The two special fittings made for this purpose are the sisson insertable joint and the kaffer tee.

Sisson Insertable Joint

The sisson insertable joint is an extra long bell with a lead ring, figure 13-1. To install an insertable joint, place the lead ring on the length of pipe that goes into the joint. Next, push the long bell onto the pipe over the lead ring. Place the bead end of the sisson joint into the other fitting and make the joint.

The sisson insertable joint is not recommended for stacks because it does not support the pipe weight above it. It is made in 2, 3, 4, and 6-inch sizes.

Kaffer Tee

The kaffer tee is similar to a sanitary tee or a Y branch, figure 13-2. It is the most efficient fitting for stacks.

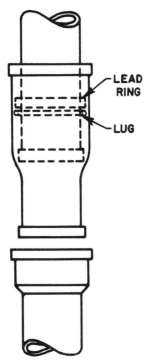

Fig. 13-1 Sisson insertable joint

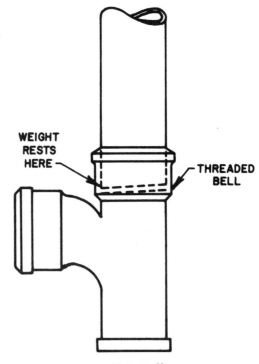

Fig. 13-2 Kaffer tee

The bell on the straight run of the fitting is threaded so it may be removed. To install a kaffer tee, remove the bell and, after cutting the stack, cut a space for the exact length of the fitting. This is important because the upper section of the stack must rest on the tee and be firmly supported. Next, slide the bell onto the upper pipe above the fitting, insert the tee, and replace the bell. Caulk the joints.

Kaffer tees range in sizes from 2 to 6 inches. They also come in wye patterns.

REVIEW QUESTIONS

Multiple Choice

Select the best answer for each question.

1. The primary use for insertable joints is

    a. to provide cleanouts in an existing system.
    b. to provide a flexible joint.
    c. to aid alterations and repair work.
    d. in sink wastes.

2. The _____ is not recommended for stacks.

    a. sisson joint
    b. kaffer tee
    c. sanitary tee
    d. Boston cleanout

3. With _____, the pipe above the fitting is firmly supported.

    a. the sisson joint
    b. the kaffer tee
    c. either a or b
    d. none of the above

4. When installing a sisson joint, the first thing to do is

    a. pack oakum into the bell.
    b. unscrew the hub.
    c. caulk the joint.
    d. slip the lead ring up the spigot end.

5. Kaffer fittings are made in sizes of

    a. 1 and 2 inches.
    b. 2 and 3 inches.
    c. 3 and 4 inches.
    d. 2 to 6 inches.

6. The sisson joint is made in sizes of

    a. 1 and 2 inches.
    b. 2 and 3 inches.
    c. 2, 3, and 4 inches.
    d. 2, 3, 4, and 6 inches.

7. The _____ has a threaded bell.

    a. kaffer fitting
    b. sisson joint
    c. neoprene joint
    d. ball seal

8. The most efficient fitting for stacks is the

    a. sisson joint.
    b. kaffer tee.
    c. sanitary tee.
    d. neoprene joint.

9. Why must the stack be cut to the exact length of the fitting when installing a kaffer tee?

    a. This provides a flexible joint.

    b. This allows the lead ring to be slipped on the spigot end.

    c. This allows the belt to be removed.

    d. The upper stack rests on the tee and must be supported.

SECTION 3

Cold Water Supply

UNIT 14 GLOBE, GATE, AND CHECK VALVES

OBJECTIVES

After studying this unit, the student will be able to:

- identify the kinds and sizes of globe, gate, and check valves.

- explain how these valves operate and where and how they are used.

Many types of valves are made for various purposes. They are made for water, steam, oil, gasoline, and other types of pipelines. They are also made to withstand certain temperatures and pressures.

GLOBE VALVES

Globe valves are used on water, air, gas, oil, or steam lines. Essentially they are control valves. They may be either fully opened or partially closed to regulate the flow.

Globe valves are made of brass, bronze, steel, or cast iron. Brass and bronze valves are made in 1/8 to 3-inch sizes. They can

withstand pressures of 125, 150, 250, and 300 psi. Steel and cast-iron valves are made in 2 1/2 to 12-inch sizes and can withstand pressures of 125 psi. The 2 1/2 to 6-inch sizes can also withstand pressures of 1500 psi. Brass-trimmed iron body valves are made in 2 to 16-inch sizes.

A disc stops the flow in a globe valve. There are four types, figure 14-1:

- The *conventional disc* closes against a beveled seat.

- The *composition disc* has a wide choice of disc material for hot or cold water, air, oil, or steam.

CONVENTIONAL

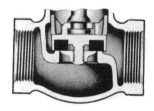
COMPOSITION

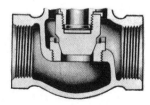
PLUG-TYPE

Fig. 14-1 Globe valves

- The *needle valve type* is used for fine throttling control. It is usually placed on gasoline or oil lines.

- The *plug-type valve* has a broad contact between the plug and the seat. The plug and seat are made of a special alloy steel and may be used where conditions are severe, such as throttling steam.

The better valves have renewable seats, figure 14-2. All these valves operate on the principle of the screw to raise or lower the disc.

Globe valves are made in straight, angle, wye, and radiator patterns. The bonnet screws into the body on some. Others have a yoke under the body or a loose nut. Larger valves are bolted. The friction in globe valves is about 60 times greater than in gate valves.

Globe valves may be joined to pipe by screwed, welded, sweated, or flanged joints.

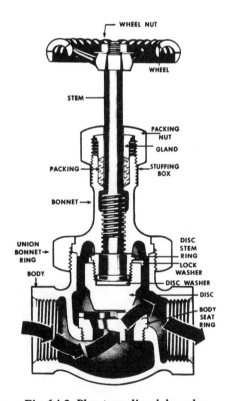

Fig. 14-2 Plug-type disc globe valve

When making a screwed joint, the wrench is placed on the valve end into which the pipe is inserted. This prevents strain on the valve. On a welded or sweated joint, the stem is removed, or at least backed off, to avoid damage to the disc.

On water systems, the valve is installed so that the incoming pressure is under the seat. This allows the stem packing to be repaired without shutting down the entire system. A mechanic attempting to replace the packing could be burned if the valve is not installed this way.

On some steam systems, the valves are installed so the pressure is over the seat. This keeps the stem from contracting as it cools and prevents the valve from opening.

On systems where it may be necessary to completely drain the lines, valves on horizontal lines must be installed with the stems horizontal. On vertical globe valves, the seat forms a dam which may only allow the line to be partially drained.

Most globe valves may be repacked at the stem. This is done by removing the packing nut, gland, and old packing and then reassembling it with new packing. The nut should not be tightened too much as the stem may be scarred.

GATE VALVES

The gate valve, also called a *stop valve*, operates on the screw principle. The wedge-shaped gate moves up and down at right angles to the path of flow between two perpendicular rings. When seated against these rings, it shuts off the flow.

The gate valve causes no more friction than a straight pipe. It is used for liquid, pump, and main lines where maximum flow is required. Gate valves should not be used for throttling as they will chatter and vibrate. When throttled, the seat may be cut by wire drawing.

Gate valves are made with a rising or nonrising stem, figure 14-3. The nonrising stem type has a left-hand thread upon which only the disc is raised or lowered.

Gate valves are also made with three types of gates. The solid-wedge gate with angle seat is recommended for steam, water, oil, air, or gas lines. It may be installed in any position, but must not be strained. Strain may change the fixed faces of the seat and cause leaks. The split-wedge and the double-disc gate valves must be installed in a vertical position. Otherwise, sediment may interfere with their operation. They may chatter if placed on lines having a high velocity.

Three types of bonnets are used on gate valves. The inside screw bonnet is used for low pressure. The union ring bonnet, figure 14-4, reinforces to the body and is ideal for frequent inspection. The bolted bonnet is used on cast-iron and steel valves for high pressure.

Valves may be attached to pipe by screwed, sweated, flanged, or welded joints. Unions or flanges should be placed between the valves and equipment. The pipe should be well supported and allow for expansion to prevent sagging or strains.

Gate valves are made in bronze, ranging in sizes from 1/4 to 3 inches. Standard bronze valves are made to withstand 125, 150, 175, or 250 psi of steam pressure. However, the pressure recommended varies inversely as the temperature rises. Larger sizes are made of cast iron and steel.

CHECK VALVES

Check valves are installed in pipelines to prevent fluids from flowing in the wrong direction. They are made in two basic types, swing check and lift check, each having several different patterns. Check valves are made of brass in the smaller sizes (1/8 to 2 inches) and cast iron in the larger sizes (3 to 6 inches).

In the *swing check valve*, figure 14-5, the flow is straight through a tilted seat. When the flow stops or reverses, the swing disc falls against the seat and closes the valve. This valve causes very little friction and is used for low to moderate pressures.

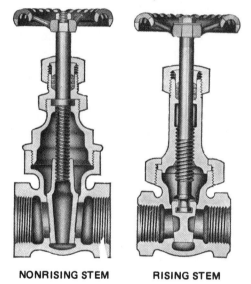

NONRISING STEM   **RISING STEM**

Fig. 14-3 Types of gate valves

Fig. 14-4 Union ring bonnet

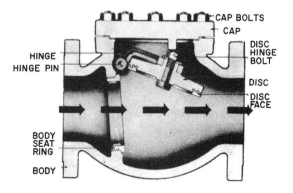

Fig. 14-5 Swing check valve

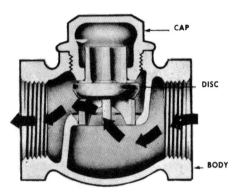

Fig. 14-6 Lift check valve

Swing check valves are used on return-circulating systems, pumps, waterlines, and main house drains. When installed on return-circulating lines, they prevent cold water from being drawn from the last fixture.

The flow through a *lift check valve* is similar to that of a globe valve, figure 14-6. It causes more friction than the swing-check valve. It is generally used on the same systems as the globe valves. It holds tighter and can withstand higher pressure. It is used for water, steam, air, gas, and vapor systems. Lift check valves are made in the same sizes as swing check valves.

Lift check valves are made in globe, angle, vertical, and horizontal patterns. Other patterns include the ball disc (a stainless steel ball) and the cushioned disc (a composition disc).

Lift check valves are used on mixing valves to prevent either hot or cold water from passing to the opposite line whenever there is a drop of pressure in one of the lines. They are also used to prevent polluted water from being siphoned back into the water system whenever there is a pressure failure.

Care must be taken when installing swing and lift check valves. They operate by gravity and will not work properly if installed at the wrong angle.

Other Check Valves

This is only a partial list of the types of check valves. Two other commonly used types are the *ball check valve* and the *vertical check valve*. Both are used on domestic water systems where a well is the water source.

REVIEW QUESTIONS

Multiple Choice

Select the best answer for each question.

1. A globe valve may be used as a control valve because
   a. it will allow flow in one direction only.
   b. it may be operated partially open.
   .c. it has a pressure side.
   d. the valve stem does not rise.

2. The globe valve uses a _____ to stop the flow.

   a. gasket   c. gate
   b. flange   d. disc

3. The gate valve uses a _____ to stop the flow.

   a. disc   c. stainless steel ball
   b. plug   d. gate

4. The globe valve slows down the flow about _____ as much as a gate valve.

   a. one-half
   b. 2 times
   c. 60 times
   d. 120 times

5. When tightening valves onto a pipe, the wrench is placed

   a. on the valve opening that has the pipe in it.
   b. on the bonnet assembly.
   c. on the valve opening opposite the pipe.
   d. none of the above.

6. The two types of gate valves are the

   a. disc and plug.
   b. rising and nonrising stem.
   c. left hand and right hand.
   d. none of the above.

7. On globe valves, the pressure is usually

   a. on the left side.
   b. against the stem.
   c. on the right side.
   d. under the seat.

8. The purpose of a check valve is to

   a. stop the flow of liquids or gases when shut off.
   b. provide a release for excessive pressure.
   c. allow flow in one direction only.
   d. allow visual inspection of the flow.

9. What causes the swing and lift check valve to operate?

   a. Gravity
   b. Atmospheric pressure
   c. The screw principle
   d. Vacuum

10. Which type of bonnet is best to use if the valve will be disassembled often?

    a. Solid plug type
    b. Inside screw
    c. Union ring
    d. Diaphragm

11. Which kind of valve may be used to throttle the flow?

    a. Globe
    b. Gate (rising stem)
    c. Gate (nonrising stem)
    d. Check

12. Which type of valve restricts the flow the least?

    a. Globe
    b. Check
    c. Relief
    d. Gate

13. Nonrising-stem gate valves are generally found in

    a. steam boilers.
    b. tight places.
    c. toilet tanks.
    d. water tanks.

14. What causes the globe valve to open and close?

    a. The screw principle
    b. Atmospheric pressure
    c. The lever principle
    d. Gravity

15. What happens when a check valve is not placed at the right angle?

    a. It eventually corrodes shut.
    b. It causes a pressure buildup.
    c. It does not operate.
    d. None of the above.

UNIT 15 WATER METERS, CURB BOXES, AND STRAINERS

OBJECTIVES

After studying this unit, the student will be able to:

- describe how water meters, curb boxes, and strainers operate.
- read water meters and curb boxes.

WATER METERS

Water meters are instruments for measuring the quantity of water used in buildings. They are placed on the main waterline before any branches are connected. In small dwellings, they are sometimes installed in a pit at the curb so they can be read at any time, figure 15-1.

Meters are installed in an upright position and where they may be easily read. Since they cause considerable friction, a meter one size larger than the pipe is sometimes used. Meters must be protected from freezing. If freezing is still possible, the meter should have a breakable bottom. This protects the vital meter parts.

A valve is placed within 12 inches of the meter on the pressure side on the waterpipe.

On 2-inch and large-size meters, a valve must be placed on each side along with a tee for testing. This tee is placed on the house side of the meter. A bypass is sometimes installed on large meters to permit uninterrupted service in case repairs must be made to the meter. Dirt, grit, and pipe dope should be blown out of the line before meters are installed. An arrow on the meter indicates the direction of flow. In new buildings, a temporary connection is installed in place of the meter to prevent damage to the meter.

Ordinarily, pipes are strong enough to support small meters, but large ones require masonry support.

Hot water swells the hard-rubber disc and prevents proper registering. It is therefore prevented from backing through the disc

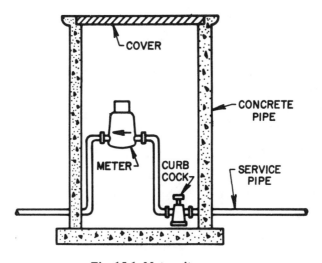

Fig. 15-1 Meter pit

meters. This is done by installing a check valve and a safety valve. A bypass may be installed to allow hot water to pass around the meter, figure 15-2. However, the use of the bypass is regulated by some codes.

Some meters register in gallons; others register in cubic feet. Some have a straight reading dial; others have circular reading dials, figure 15-3. To read a circular face dial, start at the highest numbered dial and write the last number that the hand has passed. Read the next lower dial and place this number to the right of the previous number. Continue until the lowest dial is read. This is the number of gallons or cubic feet used. A previous reading subtracted from this reading will give the amount of water used since that reading. The one-foot dial shown on the right in figure 15-3 is used for testing the meter accuracy but not for recording.

CURB BOXES

In order to turn the curb stop handle, an extension curb box is installed. The box is placed directly over the curb cock so the stop key will easily engage the tee handle, figure 15-4.

The box is set on a brick or stone support to prevent settling. If the valve is not in the center or not in an upright position, it will be difficult to engage the handle with a stop key, figure 15-5.

The curb stop is placed about 4 feet deep to prevent freezing. The top of the box is flush with the pavement for pedestrian safety. If the box is covered, its location is determined by locating the water main and tracing it back to the box. A flashlight or a mirror in the sun may be used to see the bottom of a curb box.

STRAINERS

In passing through water mains and pipes, water picks up grit, rust, scale, and

Fig. 15-2 Bypass

READING - 96872 CU. FT.

READING - 41873 CU. FT.

CIRCULAR READING DIAL

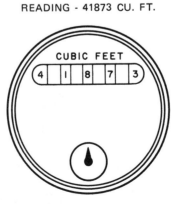

STRAIGHT READING DIAL

Fig. 15-3 Reading a water meter

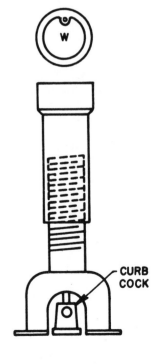

Fig. 15-4 Curb box

other harmful materials. If this grit is caught under the washer of a valve or faucet, it may destroy the washer and cut the seat. If it is caught in a regulating valve, it prevents the valve from controlling the water pressure. This, in turn, may cause a tank or piece of equipment to fail and do considerable damage. To prevent this, strainers are installed in the lines ahead of such valves and equipment.

Manufacturers of regulating valves, realizing their value, place strainers on their equipment. Strainers must always be placed on steam-reducing valves. Failure to do so

may cause an explosion. The proper location for a strainer is between the shutoff valve and the regulating valve or piece of equipment which is to be protected, figure 15-6.

Good strainers have an effective area twice the pipe size area. The strainer should be the same size as the pipe.

On large valves, a bypass is installed so that water or steam may be temporarily supplied while repairs to the regulating valve are being made. Figure 15-7 shows a regulating valve, strainer, and bypass.

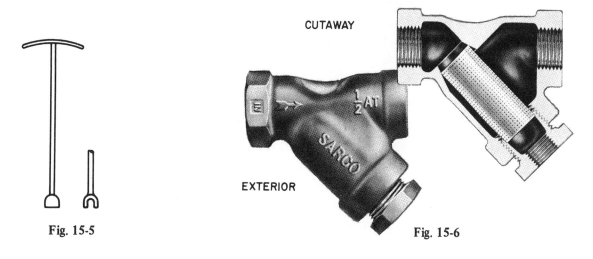

Fig. 15-5

Fig. 15-6

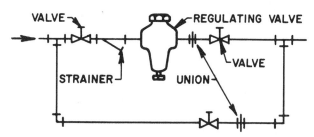

Fig. 15-7 Regulating valve, strainer, and bypass

REVIEW QUESTIONS

Multiple Choice

Select the best answer for each question.

   1. The purpose of a water meter is to measure the
      a. quality of water.   c. temperature of water.
      b. quantity of water.   d. pressure of water.

2. What should be done to the service pipe before a meter is installed?

    a. A curb cock should be installed.
    b. It should be blown out.
    c. It should be coated with bituminous paint.
    d. It should be wiped clean.

3. The purpose of a bypass on a water meter is to

    a. reduce water costs.
    b. provide additional support for the water meter.
    c. allow expanding hot water to back up into the service pipe without going through the meter.
    d. allow removal of the water meter.

4. The curb cock must be installed exactly upright to

    a. make it easier to put a curb key on it.
    b. satisfy the water company's requirements.
    c. prevent trapping the water service.
    d. permit periodic cleaning.

5. Dirt and grit is prevented from damaging valve seats by using a

    a. sand filter.   c. bent wire.
    b. settling tank.   d. strainer.

6. Larger regulating valves are installed with a bypass because

    a. the valve must be back-flushed from time to time.
    b. the supply must be maintained while the valve is being repaired.
    c. the bypass will be used when the strainer becomes blocked.
    d. none of the above.

7. Whenever possible, meters are placed inside to

    a. make servicing easier.   c. make installation easier.
    b. make meter reading easier.   d. prevent freezing.

8. To determine the amount of water used from reading to reading, the numbers are

    a. divided.   c. added.
    b. multiplied.   d. subtracted.

9. With respect to the meter, at least one valve is placed

    a. at the curb.
    b. just upstream of the meter.
    c. just downstream of the meter.
    d. between tees.

10. In relation to the pipe, the strainer should be

    a. one size larger.   c. the same size.
    b. two sizes larger.   d. one size smaller.

UNIT 16 PRESSURE-REDUCING VALVES AND THE BYPASS

OBJECTIVE

After studying this unit, the student will be able to:

- discuss the uses and purposes of pressure-reducing valves and the bypass.

PRESSURE-REDUCING VALVES

Pressure-reducing valves are installed to protect water supply systems from excessive pressure (over 60 psi). The valve is located near the point of entrance when the whole system is to be protected. In large buildings they are used in certain zones or floors. They are also used to protect special fixtures or equipment. They must be placed on supplies to heating systems in small buildings. This is because heating boilers and radiators are guaranteed to stand only 30 psi of pressure.

When pressure-reducing valves are installed on the supply to any heating device, a safety valve must also be installed. The safety valve is placed between the pressure-reducing valve and the heater to guard against an explosion.

Pressure-reducing valves for water supplies are the diaphragm type, figure 16-1. Diaphragm valves work on the principle that low pressure on a large area (the diaphragm) will overcome a high pressure on a small area (the valve disc).

The valve consists of a body with the seat facing down. The stem, with a washer on the bottom, extends upward and is attached to the diaphragm. As the pressure in the house side reaches the desired number of pounds, the pressure pushes up on the diaphragm. This pulls the washer against the seat and closes the valve. The spring is placed on top of the diaphragm to assist in opening the valve and for smoother operation.

The pressure at which the valve closes may be increased by tightening an adjusting screw on top of the valve. The washer is renewed by removing the cap at the bottom of the valve. The diaphragm may be renewed by removing the bolts around the rim.

Better results are obtained when the valve operates completely open. The line is always blown out before connecting the reducing valve. A strainer is always placed ahead of the reducing valve. Large valves are bypassed.

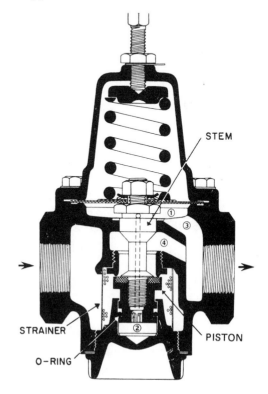

Fig. 16-1 Diaphragm pressure-reducing valve

THE BYPASS

When reducing or thermostatic-control valves are installed on steam or waterlines, they usually have a bypass around the valve. A *bypass* permits water or steam to be used temporarily when the reducing or thermostatic valves are out of order, figure 16-2.

If the reducing valve requires repairs, the gate valves in the 2-inch line are closed, and the 1 1/4-inch globe valve on the bypass is opened. This provides a temporary supply. Watch gauge B on the low-pressure side to prevent excessive water or steam pressure.

When there is a thermostatic-regulating valve where the steam pressure is high, the corresponding steam temperature may heat the water to a dangerous degree. A safety valve must be placed on the low-pressure side of all regulating valves. This prevents pressure due to expansion from backing through such a valve and causing an explosion.

A gauge should be placed on each side of the valve so that both pressures may be readily seen. A strainer must also be placed in front of a reducing valve to prevent chips or rust scale from entering. If chips get caught beneath the seat washer, the valve will not shut off tightly and pressure will increase on the low-pressure side. The safety valve should be set about 25 percent higher than the desired low pressure. Unions are placed on each side of the reducing valve to make its removal easier.

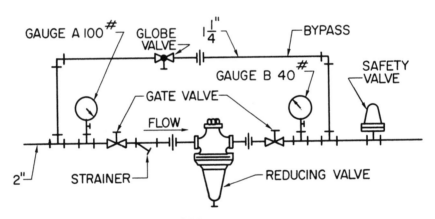

Fig. 16-2 Bypass

REVIEW QUESTIONS

Multiple Choice

Select the best answer for each question.

1. The purpose of a bypass is to

    a. support the pipe when the reducing valve is removed.

    b. mix high-pressure steam with low-pressure steam.

    c. provide manual operation of the system when the pressure-reducing valve is removed.

    d. prevent dirt from accumulating.

2. The principle on which the pressure-reducing valve works is

   a. a low pressure over a large area can overcome a high pressure over a small area.
   b. a high pressure over a small area can overcome a low pressure over a large area.
   c. a heavy spring can overcome atmospheric pressure.
   d. none of the above.

3. Pressure-reducing valves used to supply water boilers are the

   a. spring type.   c. diaphragm type.
   b. float type.   d. lever type.

4. The outlet pressure on a reducing valve can be increased by

   a. placing a strainer in the line.
   b. adjusting the handwheel.
   c. loosening the adjustment screw.
   d. tightening the adjustment screw.

5. A safety valve must be installed on the low-pressure side of a reducing valve because

   a. increased temperatures are caused by a reduction in pressure.
   b. the reducing valve may develop excessively high pressures.
   c. pressure must be regulated by hand when the valve is removed.
   d. the valve will not allow expanded volume to back up.

6. To replace a defective washer on a reducing valve,

   a. the bolts around the top must be removed.
   b. the strainer must be removed.
   c. the cap must be removed from the bottom.
   d. the plug must be removed.

7. To replace a defective diaphragm on a reducing valve,

   a. the bolts around the top must be removed.
   b. the strainer must be removed.
   c. the cap must be removed from the bottom.
   d. a wire must be inserted into the inlet.

8. A strainer is installed on the high-pressure side to

   a. keep dirt out of the heating equipment.
   b. keep rust and chips from damaging the reducing valve.
   c. filter mineral deposits from the water.
   d. flush the strainer completely.

9. If the low-pressure side is 40 psi, the relief valve pressure should be

   a. 50 psi.   c. 80 psi.
   b. 65 psi.   d. 100 psi.

10. The purpose of the unions on both sides of the valve is to

    a. adjust the pressure.
    b. make a swing joint.
    c. provide a place to bleed-off scale and rust.
    d. make it easier to remove.

UNIT 17 HOUSE SERVICE PIPE
AND BASEMENT WATER MAINS

OBJECTIVES

After studying this unit, the student will be able to:

- discuss the purpose and use of house service pipe and basement water mains.

- identify the cause of sweating pipes and explain how to correct it.

HOUSE SERVICE PIPE

The house service pipe carries water from the street main into the building. A *corporation ferrule* is inserted in the street main by the city water department. The ferrule consists of a ground key cock with a special thread by which it is screwed into the street water main. A coupling to which the service pipe is attached is located on the opposite end.

The size of the house service pipe is determined by the number of fixtures to be supplied and the size of the water system. Small house service pipes are at least 3/4-inch type-K soft copper tubing.

The service pipe must be laid about 4 feet deep, depending on the location, to prevent freezing. It must also be supported at the main to prevent breaking, figure 17-1. The round way, inverted curb cock is the best type to use. The opening is as large as the pipe and will require no repairs for years.

Buildings requiring a pipe larger than 2 inches usually have cast-iron service pipe with a caulked joint. Plain, not tarred, oakum must be used in the joint, or the water will carry a taste of tar.

From the tapping in the street main, the house service line must run straight to the curb cock. From the curb cock, the

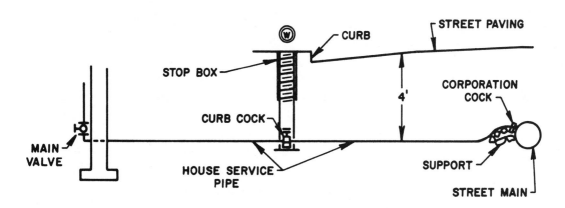

Fig. 17-1 Supporting the service pipe

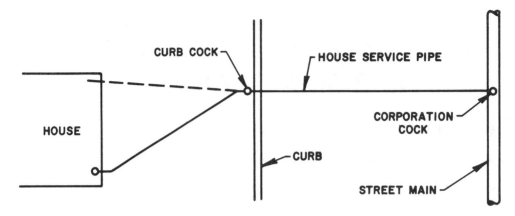

Fig. 17-2

pipe may run in any direction on the individual's property, figure 17-2.

To install a new water service pipe, the master plumber obtains a ferrule and a permit from the local government to open the street. The plumber assumes the responsibility of protecting the public. Signs are placed on fences and on any open trench during the day, and lanterns are placed at night.

As indicated in figures 17-1 and 17-2, there are three places to shut off the water. The corporation ferrule may be shut off with a wrench after the street is opened. The curb cock may be shut off with a stop key lowered down through the stop box. The valve in the basement may be turned by hand.

BASEMENT WATER MAINS

The basement water main is that part of the water supply system which extends through the basement of a building. The branches to all fixtures are connected to the basement main.

The basement main is attached to the house service pipe where it enters the building. Brass, copper, galvanized iron, or steel pipe and fittings may be used. It should run directly with as few bends as possible to avoid friction.

The basement main must be large enough to supply all the fixtures that may be used at one time. There should be at least 15 psi of pressure at the highest fixture. Failure to provide this may cause back-siphoning, particularly in high buildings, and may also pollute the water supply.

Note: The water supply of all buildings is metered to prevent the waste of water.

If galvanized iron pipe is used, basement mains in small buildings should be 1 inch minimum to allow for corrosion. If the mains are made of copper, they may be 3/4 inch.

A properly designed system has a main shutoff valve at the point of entrance. One is for the hot water tank, and one is for each fixture or riser. Compression valves with waste are used in small buildings. Larger systems have globe valves. In the latter case, other valves are placed at the bottom of risers for draining purposes, figure 17-3.

Basement mains should be supported every 10 feet. Reznor hooks are used for small pipes. Pipe hangers are used for large mains.

SWEATING OF PIPES

Moisture appearing and dropping from the surface of cold water, ice water, or brine pipes is caused by warm, damp air condensing on the cold pipe. The drops appear along the bottom of the pipe as shown in figure 17-4. This sweating should be avoided since moisture dropping from pipes can damage floor coverings, furniture, or merchandise.

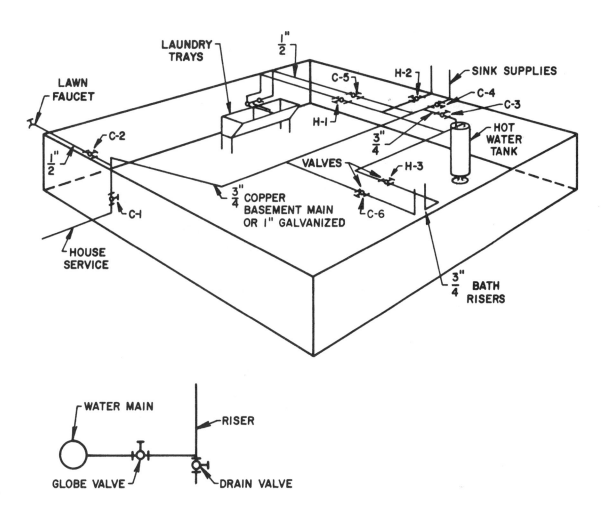

Fig. 17-3

Two conditions must be present to cause sweating of pipes. First, the air must contain a high percentage of moisture. Second, the pipe must be cold. This usually happens in the summer during hot, humid weather. It may also take place in a warm building in which considerable water and steam are used, such as in laundries, dye houses, or dairies. In these buildings, the water pipes are kept cold by continuous use. Under these conditions, sweating occurs. In a home, a leaking faucet or tank ball cock will cause the cold water pipes to sweat.

To prevent sweating of pipes, cover the pipes to prevent the moisture-laden air from coming in contact with the pipes. This is done by covering the pipes and fittings with hair felt or other special covering and painting the outside of the coverings with several coats of paint. This must be done carefully or moisture may penetrate.

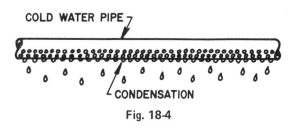

Fig. 18-4

Fig. 17-4 Sweating pipe

REVIEW QUESTIONS

Multiple Choice

Select the best answer for each question.

1. The position of the tap in the street is determined by
   a. visual observation.
   b. using a magnetometer.
   c. measuring out from the building line.
   d. the position of the curb box.

2. What kind of copper tubing is used for a house service pipe?
   a. Hard DWV tubing
   b. K-type copper tubing
   c. L-type copper tubing
   d. None of the above

3. A corporation ferrule is located
   a. on every compression fitting.
   b. in a factory building.
   c. in a basement.
   d. under a street.

4. Small house-service pipes must be at least
   a. 1/2 inch.
   b. 3/4 inch.
   c. 1 inch.
   d. 1 1/4 inches.

5. To shut off the corporation ferrule, the first thing to do is
   a. shut off the meter valve.
   b. take the cover off the curb box.
   c. dig up the street.
   d. open all the valves in the house.

6. To shut off the curb cock, a _____ is required.
   a. stop key
   b. wrench
   c. plumbing permit
   d. vent plunger

7. There must be at least _____ pounds per square inch available at the highest and farthest fixture.
   a. 4
   b. 8
   c. 12
   d. 15

8. What type of hanger is most likely to be found on a 3/4-inch basement water main?
   a. Clevis hanger
   b. Renzor hook
   c. Spring
   d. Belted ring

9. The water drops on sweating pipes come from
   a. moisture in the air.
   b. small leaks within the pipe.
   c. vaporization of air.
   d. heat.

10. To prevent pipes from sweating, they must be
    a. watertight.
    b. insulated.
    c. painted with waterproof paint.
    d. dried out.

UNIT 18 PRIVATE WATER SYSTEMS

OBJECTIVES

After studying this unit, the student will be able to:

- describe the parts of a typical private water supply system.
- explain how the parts of the system work together.

PRIVATE WATER SYSTEMS

Private water supply systems may use springs, creeks, cisterns, and other sources for their water supply. The typical system uses a cased well for its water source. If properly constructed, this is the most reliable, pollution-free water supply. The parts of the private system are the:

- Pump
- Pressure tank
- Pressure switch
- Foot valve
- Air volume control

How the System Works

The pump of a well either lifts or pushes water from the well into a tank until the desired pressure is reached (40 to 45 psi), figure 18-1. A pressure-actuated electrical switch turns the pump on and off. The switch might be adjusted to turn on the pump when the pressure in the tank drops to 20 psi, and then turn it off when the pressure climbs to 40 psi.

Since water is incompressible, the pressure tank operates with a cushion of air trapped in the top. If the tank is completely filled with water, the pressure in the tank will drop immediately when a tap is opened.

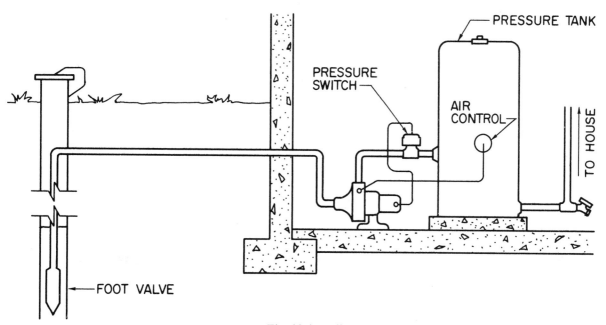

Fig. 18-1 Well

The pump will then turn on to restore pressure. Because there is no air cushion, the pressure will build up immediately, and the pump will shut off again. This will repeat again and again before a small container fills with water.

Air is compressible. This property enables it to expand without a rapid pressure loss. It is important to maintain an air cushion in the top of the tank. To maintain this air cushion a device, called an *air control*, is used. A foot valve is placed at the end of the pipe submerged in the well. It prevents the water being drawn up the well pipe from falling back down into the well when the pump turns off. If this check valve is not there, the pump will lose its *prime* when it stops running. When a pump is primed, it is full of water and has no air in it. Water pumps will not operate when they are air-bound.

PUMPS

There are several types of water pumps and pumping systems. A pump may be a jet type, submersible, or reciprocating. These three pumps may be further classified for either a shallow well or deep well.

A shallow-well pump is limited to a theoretical depth of 35 feet. In practice, because of friction loss and pumping efficiency, depths of 25 feet are usually the limit for this type. The shallow-well pump uses atmospheric pressure (14.7 psi) to lift the water. Sucking water up a straw is a good comparison.

The more common deep well-pump pushes the water up the well pipe to the tank. The submersible deep-well pump accomplishes this by being lowered into the water in the well. It is sealed so that water cannot enter the electric motor. The deep-well jet pump uses a two-pipe system within the well to push and draw the water to the surface,

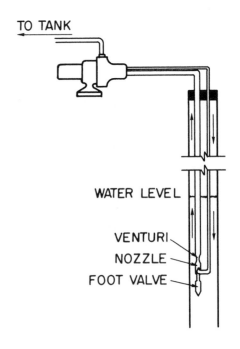

Fig. 18-2 Deep-well jet pump

figure 18-2. The jet pump diverts some of its water back down the well. This water is forced through a *venturi*, a narrow opening in the jet body, to create a low pressure. This low pressure assists the pump in lifting the water up the pipe.

The reciprocating or piston-type pump must use a long pump rod extending to a piston which is submerged in the well water. Because of the complexity of this system, the reciprocating pump is usually limited to shallow-well applications for the home.

THE PRESSURE SWITCH

The pressure switch is mounted on the tank or in the discharge pipe from the pump. It uses the existing pressure to activate an electrical switch. In a common application, it *makes* or turns the pump on at 20 psi, and *breaks* or turns the pump off at 40 psi.

AIR CONTROL VALVES

The job of the air control valve is to add a small amount of air to the pressure tank each time the pump cycles. Without air control the water would slowly absorb the air cushion in the tank.

There are several methods of admitting air to the pressure tank. Most use the negative pressure on the suction side of the pump to draw air in, figure 18-3. The air control valve, or *air charger*, must add air only when the tank requires it.

THE PRESSURE TANK

The pressure tank may simply be a cylindrical tank designed to handle the maximum working pressures of the system. Some tanks have a plastic foam float which rides on the water surface in the tank. The float separates the air and the water. This prevents the water from absorbing the air. Some tanks separate the air and water with a rubber diaphragm or air bladder.

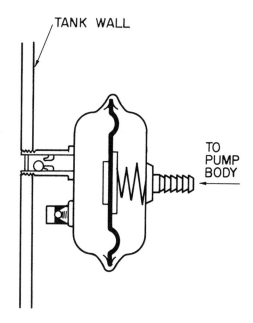

Fig. 18-3 Air control

REVIEW QUESTIONS

Multiple Choice

Select the best answer for each question.

1. Which of the following is not a possible source of water for a private water system?
   a. Wells
   b. Springs
   c. Cisterns
   d. Cesspools

2. The practical depth limit of a shallow-well pump is most nearly
   a. 12 feet.
   b. 25 feet.
   c. 34 feet.
   d. 124 feet.

3. The switch which turns the pump on and off is actuated by
   a. pressure.
   b. temperature.
   c. weight.
   d. vacuum.

4. It is important to maintain _____ in the pressure tank.
   a. a rubber seal
   b. a foot valve
   c. an air cushion
   d. a flood seal

5. If the pump cannot maintain its prime, the _____ may be defective.
   a. well casing
   b. foot valve
   c. pressure switch
   d. pressure tank

6. The deep-well jet pump uses _____ to assist in lifting the water.

    a. a venturi
    b. an air control

    c. a cylinder
    d. an air pump

7. A foot valve is a type of

    a. pressure switch.
    b. depth control.

    c. well seal.
    d. check valve.

8. If there is no air control, the water in the pressure tank will

    a. rust the tank.
    b. become contaminated.

    c. leak out.
    d. absorb the air.

9. A deep-well reciprocating pump uses a _____ in the well casing.

    a. long pump rod
    b. two-pipe jet

    c. bronze foot valve
    d. soft rubber seal

10. The safest water source for a private water system is

    a. a spring.
    b. a cistern.

    c. a deep well.
    d. a river.

UNIT 19 WATER PRESSURE AND HYDROSTATIC PRESSURE

OBJECTIVES

After studying this unit, the student will be able to:

- explain water and hydrostatic pressure.
- determine water pressure and hydrostatic pressure.
- discuss the importance of air chambers.

WATER PRESSURE

To understand water pressure, consider the weight of a cubic foot of water at average temperature to be 62.5 pounds. Since pressure is usually given in pounds per square inch, consider the cubic foot as 144 columns, 1 square inch in cross section and 12 inches high. The weight of each column is 62.5 pounds divided by 144, or .434 pound.

If this column is placed vertically and a gauge is attached to the bottom, it shows that the weight and pressure (.434) are equal. However, this is only true when the surface area is 1 square inch. Pressure per square inch (psi) is computed by the vertical height above a certain point and is exerted outward in all directions.

A gauge attached to the bottom of a cubic foot of water shows no more pressure than the 1-inch column described. The area above 1 inch does not affect the pressure; neither does the shape of the vessel.

If the 144 one-inch square columns shown in figure 19-1 are placed vertically above each other, they will exert a pressure of .434 × 144. This equals 62.496 or 62.5 psi. Note that this is also the weight of a cubic foot.

Pressure is affected by friction in moving bodies of water. To determine the pressure per square inch at the bottom of a tank, multiply the *head* or height of the water above the bottom by .434. For example, the pressure at the bottom of a tank that is 8 feet deep is .434 × 8, or 3.472 psi.

Study the columns in figure 19-2, page 72. Columns 2, 3, 4, and 5 are different sizes and shapes. However, they are each 12 inches high. By opening each valve separately, the gauge will register .434 psi, the pressure exerted by a column 1 foot high.

Column 1 is 4 feet high. Therefore, the pressure on the gauge would register .434 x 4, or 1.736 psi.

HYDROSTATIC PRESSURE

Pressure on any part of an enclosed liquid is exerted uniformly in all directions. The pressure acts with equal force upon all equal surfaces at right angles to them.

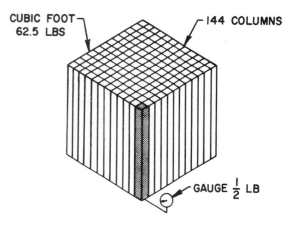

Fig. 19-1 One cubic foot

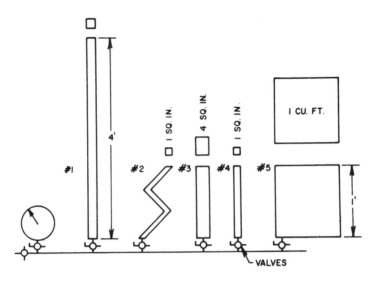

Fig. 19-2

Therefore, if a pressure of 1 psi is exerted over an area of 1 square inch upon an enclosed liquid, every square inch of the vessel is subjected to a pressure of 1 psi.

For example, a cylinder having an area of 1 square inch is connected to a cylinder having an area of 100 square inches. Each has watertight plungers. A 1-pound weight on the small cylinder will therefore support 1 pound for each inch, and 100 pounds on the large cylinder, figure 19-3.

Study the cylinder in figure 19-4 having pistons A, B, C, D, E, and F. The areas of the pistons are as follows:

> A = 100 square inches
> B = 7 square inches
> C = 1 square inch
> D = 6 square inches
> E = 8 square inches
> F = 4 square inches

Disregard the weight of the pistons and water. If a force of 5 pounds is applied to piston C whose area is one square inch, how much pressure must be applied to the other pistons to counterbalance this force?

Solution: A force of 5 pounds on piston C equals a pressure of 5 psi. Therefore, multiply the pounds per square inch by the

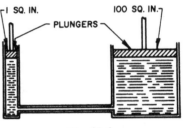

Fig. 19-3

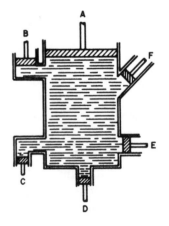

Fig. 19-4

area of each piston to obtain the number of pounds supported by those pistons. *Example:* Force A is 100 square inches times 5 pounds, or 500 psi. Using the same process, find the required force on the other pistons.

If cylinder C has an area of 8.25 square inches and a force of 150 pounds is applied to it, the forces on the other pistons are found by dividing the force by the square inches and multiplying by the area of each piston. *Solution:* 150 divided by 8.25 equals 18.182 psi. Piston D equals 18.182 pounds multiplied by 6 square inches, or 109.092 psi.

This is the principle upon which hydraulic presses, hydraulic lifts and elevators, flushometer valves, and reducing valves operate.

WATER HAMMER AND AIR CHAMBERS

Water hammer is caused by a moving body of water suddenly stopping in a pipe. Water is almost incompressible. Therefore, when a valve is closed quickly, a shock is sent through the system. This shock has been known to create pressures as high as 800 psi.

The intensity of the water hammer depends upon the volume of water, the velocity at which it is flowing, and how sudden the valve is closed. Self-closing and quick-compression faucets, valves, and pumps cause nearly all water hammer. Water hammer is also caused by suddenly opening a valve which allows water to flow under pressure into an empty pipe.

The shock and excessive pressures caused by water hammer may cause considerable damage to valves and piping. Reducing or check valves in the line make the shock less severe.

In figure 19-5, valve A is attached to pipe B containing water at 50 psi of pressure. When valve A is opened, the pressure drops to 30 psi due to friction. When valve A is closed suddenly, the pressure jumps to 680 psi. This is a tremendous shock. Instead of having only one shock, a series of shocks occur as shown in figure 19-5. Each succeeding shock becomes less intense and further apart until the water is at rest at the original 50 psi of pressure.

Air chambers are the shock absorbers of the plumbing system. They are used to relieve the shock resulting from water hammer. While water is almost incompressible, air is highly compressible. Therefore, air confined in a properly-sized pipe will act as a cushion.

Water absorbs air and, in time, the air chamber becomes waterlogged. A *petcock* is placed near the top and a valve and drain placed at the bottom to recharge the air chamber. The air chamber is placed near a quick-closing valve or at the end of a long run of pipe. Experiments have shown that

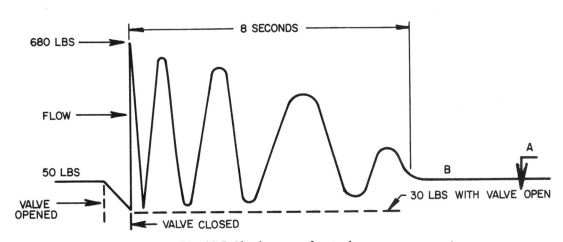

Fig. 19-5 Shock waves of water hammer

a short air chamber having a large area is better than a long air chamber with a small area. An air chamber 6 inches in diameter and 18 inches long is 27 percent more efficient than one 2 inches in diameter and 64 inches long.

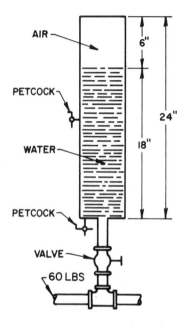

Fig. 19-6 Air chamber

When the air chamber is filled with air at atmospheric pressure (15 psi) and water at 30 psi (two atmospheres) is turned on, the air in the air chamber is reduced to one-half its original volume. Likewise, with 45 psi (three atmospheres) or with 60 psi (four atmospheres), the air is reduced to one-third and one-fourth, respectively, of its original volume.

The air chamber is placed in a vertical position to receive any air from the water main, figure 19-6. The top petcock is placed 6 inches below the estimated water level to prevent air leakage. The valve and petcock are placed at the bottom for drainage. Notice that with pressure of 60 psi (four atmospheres), the air volume is reduced one-fourth.

Since air chambers can become water-logged, making them ineffective until recharged, other devices have been developed to absorb high-pressure surges or shocks. One such device, figure 19-7, consists of an elastic, compressible material in the shape of a tube and insert. It is inserted at some point in the pipeline to serve the same purposes as an air

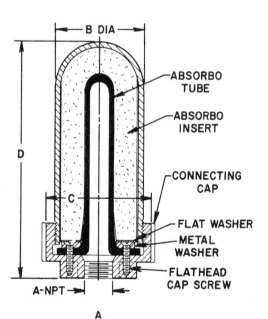

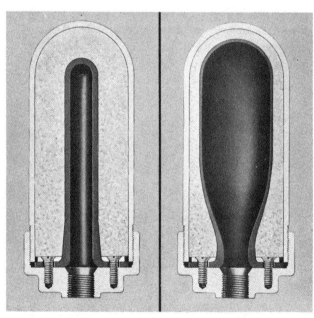

Fig. 19-7

chamber. At B, the tube and insert are in normal position before the faucet or valve is opened. The expansion of the tube against the insert as it absorbs a sudden shock is shown in C. As the shock recedes, the tube and insert return to their original position.

REVIEW QUESTIONS

Multiple Choice

Select the best answer for each question.

1. If a 1″ x 1″ x 12″ column of water weighs .434 pound, what is the weight of one cubic inch?

    a. 4.34 pounds

    b. .434 pound

    c. .0362 pound

    d. .0658 pound

2. What pressure is exerted on the bottom of a swimming pool if it is 9 feet deep?

    a. .434 psi

    b. 3.906 psi

    c. 14.63 psi

    d. 46.872 psi

3. There is 65 psi of pressure on the first floor of a building. How much pressure is on the second floor 15 feet above?

    a. 6.51 psi

    b. 9.31 psi

    c. 58.49 psi

    d. 71.51 psi

4. Hydrostatic head is

    a. the loss in pressure caused by friction.

    b. the height to the surface of a column of water.

    c. the area of the surface of water.

    d. one-half the height of a column of water.

5. In which direction is the pressure in a cylinder exerted?

    a. Towards the sides

    b. Up

    c. Down

    d. In all directions

6. A pressure gauge is installed 10 feet up from the bottom in a standing tank of water 21 feet deep. What will the gauge read?

    a. 4.34 psi

    b. 4.774 psi

    c. 5.81 psi

    d. 9.114 psi

7. If a 2-inch steel pipe with a gauge in the bottom is filled with 20 feet of water and held upright, the gauge will read 8.68 psi. If the same pipe is laid down until the top is one foot higher than the bottom, what will the gauge read?

    a. .434 psi

    b. 1.832 psi

    c. 8.246 psi

    d. 8.68 psi

8. A cylindrical tank measures 20 inches in diameter by 48 inches long and has a street pressure of 50 psi. What is the total pressure attempting to burst the boiler?

   a. 50 psi   c. 48,000 psi
   b. 86.8 psi   d. 912,236 psi

9. How much will 30 psi of pressure raise water in a pipe?

   a. 11.66 feet   c. 24.11 feet
   b. 13.02 feet   d. 69.12 feet

10. Which of the following valves is most likely to cause water hammer?

    a. An outside hose bib   c. A single-lever sink faucet
    b. A lavatory supply valve   d. A pressure relief valve

11. If an air chamber is made and installed without any valves, what is likely to happen?

    a. It will become waterlogged.   c. It will corrode shut.
    b. It will become air-bound.   d. It could not be removed.

12. What is atmospheric pressure?

    a. 4.34 psi   c. 15 psi
    b. 8.38 psi   d. 60 psi

UNIT 20 SIZING THE WATER SUPPLY SYSTEM

OBJECTIVES

After studying this unit, the student will be able to:
- explain friction loss in pipes.
- discuss water velocity.
- size pipe by the velocity method and the pressure-loss method.

FRICTION LOSS IN PIPES

To understand the loss of pressure caused by water flowing through pipe, one must consider the length and size of the pipe, the roughness of the interior, and the number and type of fittings.

In figure 20-1, tank A is filled with water. When valve E is closed, gauges B and C show the same pressure. This is the pressure exerted by the head H, or H times .434. However, when valve E is opened and water flows, gauge C shows less pressure than gauge B. This pressure loss is due to the rubbing action of the moving water against the sides of the pipe and is known as *friction loss.*

If the pipe is twice as long, the friction is twice as much. If the pipe corrodes and becomes very rough, the friction loss increases considerably. If the pipe size increases, the friction loss is reduced. This is because only that water which actually touches the side of the pipe causes friction.

Friction is proportional to the square of the velocity. Friction is increased by changes in the size of the pipe because of the eddies

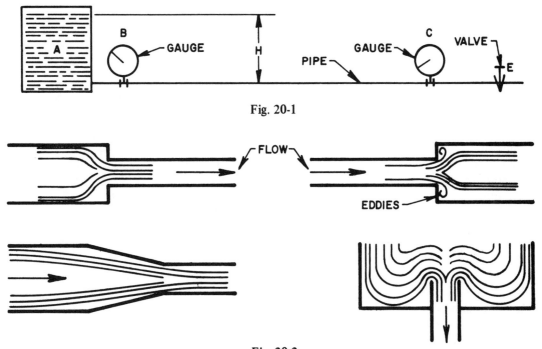

Fig. 20-1

Fig. 20-2

which are set up in the pipe as shown in figure 20-2. *Eddies* are air or water currents which move against a main current.

WATER VELOCITY

Water has a number of properties which concern the plumber. *Hardness* is the total mineral content of the water and affects the scaling of the inside pipe surfaces. The *pH* of the water means the acid balance of the water. A pH of 7 is neutral. A low pH means that the water has a high acid content. With a low pH, for instance, copper water tubing is liable to corrode quickly.

These factors can be controlled somewhat with water-softening equipment. The plumber can slow down the speed of the water in the pipe by increasing the size of the tubing. The slower the water flows, the less corrosion that takes place. Water which flows too fast:

- is noisy.
- causes water hammer.
- causes excessive wear and corrosion, especially when water is hard, acidic, or hot (over 150 degrees).

Under normal conditions velocity should be 8 feet per second (fps). If the water has a low pH (high acid), water softening equipment, quick-closing valves, or very hot water (over 150 degrees), the velocity should be 4 fps.

SIZING PIPE BY THE VELOCITY METHOD

Sizing pipe by the velocity method applies to buildings of three stories or less.

Fixture	Occupancy	Type of Supply Control	Load Values, in Water Supply Fixture Units		
			Cold	Hot	Total
Water closet	Public	Flush valve	10.		10.
Water closet	Public	Flush tank	5.		5.
Urinal	Public	1″ flush valve	10.		10.
Urinal	Public	3/4″ flush valve	5.		5.
Urinal	Public	Flush tank	3.		3.
Lavatory	Public	Faucet	1.5	1.5	2.
Bathtub	Public	Faucet	3.	3.	4.
Shower head	Public	Mixing valve	3.	3.	4.
Service sink	Offices, etc.	Faucet	2.25	2.25	3.
Kitchen sink	Hotel, restaurant	Faucet	3.	3.	4.
Drinking fountain	Offices, etc.	3/8″ valve	0.25		0.25
Water closet	Private	Flush valve	6.		6.
Water closet	Private	Flush tank	3.		3.0
Lavatory	Private	Faucet	0.75	0.75	1.
Bathtub	Private	Faucet	1.5	1.5	2.
Shower stall	Private	Mixing valve	1.5	1.5	2.
Kitchen sink	Private	Faucet	1.5	1.5	2.
Laundry trays (1 to 3)	Private	Faucet	2.25	2.25	3.
Combination fixture	Private	Faucet	2.25	2.25	3.
Dishwashing machine	Private	Automatic		1.	1.
Laundry machine (8 lbs.)	Private	Automatic	1.5	1.5	2.
Laundry machine (8 lbs.)	Public or General	Automatic	2.25	2.25	3.
Laundry machine (16 lbs.)	Public or General	Automatic	3.	3.	4.

Fig. 20-3 Load values assigned to fixtures

The water pressure available must be at least 40 psi.

Step 1. Obtain the available pressure at the main and the corrosive qualities of the water from the local authorities.

Step 2. Make a schematic drawing of the entire system. Show where the branches are located and which fixtures go with which branches. Identify all quick-closing valves, such as flushometers which require 4 feet per second piping.

Step 3. Refer to figure 20-3. Mark on the drawing the total water supply fixture units (wsfu) to each fixture and the hot-and-cold-water valves.

Step 4. Starting from the fixture outlets, mark minimum supply sizes down from figure 20-4.

Step 5. Working from the fixtures to the service pipe, size the system according to the velocity limitation tables in figure 20-5, pages 80 and 81. If the system pressure is above 40 psi, go to the nearest pipe size. If the system pressure is very close to the minimum

(40 psi), and the wsfu load falls between two pipe sizes, choose the larger.

SIZING PIPE BY THE PRESSURE-LOSS METHOD

Large buildings in area, buildings over three stories in height, buildings with a large number of outlets which run continuously (air-conditioning cooling towers etc.), and buildings with low pressures (many private water systems) should be sized by the pressure-loss method (see the *National Standard Plumbing Code*).

EXAMPLE: Determine the pipe size for a small home. It has a bathroom group (tub, toilet, and lavatory), an outside hose bib (a valve which may run continuously), a small washing machine, a laundry tray, and a kitchen sink. It has city water, and the municipal authority says the water is noncorrosive and has a minimum pressure of 55 psi.

Step 1. The pressure is above 40 psi. Because the water is noncorrosive, the system can be designed for 8 feet per second.

Fixture or Device	Size (in.)
Bathtub	1/2
Combination sink and laundry tray	1/2
Drinking fountain	3/8
Dishwashing machine (domestic)	1/2
Kitchen sink (domestic)	1/2
Kitchen sink (commercial)	3/4
Lavatory	3/8
Laundry tray (1, 2, or 3 compartments)	1/2
Shower (single head)	1/2
Sink (service, slop)	1/2
Sink (flushing rim)	3/4
Urinal (1" flush valve)	1
Urinal (3/4" flush valve)	3/4
Urinal (flush tank)	1/2
Water closet (flush tank)	3/8
Water closet (flush valve)	1
Hose bib	1/2
Wall hydrant or sill cock	1/2

Fig. 20-4 Minimum size of fixture supply pipes

Copper and Brass Pipe, Standard Pipe Size

Nominal Size (in.)	Actual I.D. (in.)	Flow (gpm) q	Velocity = 4 feet per second			Flow (gpm) q	Velocity = 8 feet per second		
			Load (wsfu) 1*	Load (wsfu) 2*	Friction (psi/100') p^3*		Load (wsfu) 1*	Load (wsfu) 2*	Friction (psi/100') p^3*
			Col. A	Col. B			Col. A	Col. B	
½	.625	3.8	1.5	—	6.8	7.6	3.7	—	24.2
¾	.822	6.6	3.0	—	5.1	13.2	8.4	—	18.0
1	1.062	11.0	6.3	—	3.7	22.0	26.4	8.0	13.3
1¼	1.368	18.3	16.8	6.4	2.8	36.6	75.0	22.7	10.0
1½	1.600	25.2	36.3	9.3	2.3	50.4	130.0	51.0	8.4
2	2.062	41.6	92.0	29.5	1.7	83.2	291.0	170.0	6.2
2½	2.500	61.2	181.0	80.0	1.4	122.4	492.0	376.0	4.9
3	3.062	92.0	335.0	209.0	1.1	184.0	842.0	807.0	3.9
4	4.000	158.0	685.0	611.0	0.8	316.0	1920.0	1920.0	2.9

Threadless Copper and Red Brass Pipe (TP)

Nominal Size (in.)	Actual I.D. (in.)	Flow (gpm) q	Load (wsfu) 1* Col. A	Load (wsfu) 2* Col. B	Friction (psi/100') p^3*	Flow (gpm) q	Load (wsfu) 1* Col. A	Load (wsfu) 2* Col. B	Friction (psi/100') p^3*
½	.710	4.9	2.0	—	5.9	9.8	5.3	—	20.8
¾	.920	8.3	4.2	—	4.4	16.6	13.2	5.7	15.5
1	1.185	13.7	9.0	—	3.3	27.4	44.0	10.5	11.7
1¼	1.530	22.9	28.9	8.3	2.4	45.8	110.0	40.0	8.5
1½	1.770	30.6	55.0	14.5	2.1	61.2	181.0	80.0	7.2
2	2.245	49.4	126.0	48.5	1.6	98.8	369.0	240.0	5.6
2½	2.745	74.0	245.0	125.0	1.3	148.0	631.0	537.0	4.4
3	3.334	109.0	421.0	305.0	1.0	218.0	1081.0	1081.0	3.5
4	4.286	180.0	816.0	774.0	0.8	360.0	2318.0	2318.0	2.6

Copper Water Tube, Type L

Nominal Size (in.)	Actual I.D. (in.)	Flow (gpm) q	Load (wsfu) 1* Col. A	Load (wsfu) 2* Col. B	Friction (psi/100') p^3*	Flow (gpm) q	Load (wsfu) 1* Col. A	Load (wsfu) 2* Col. B	Friction (psi/100') p^3*
½	.545	2.9	1.0	—	8.2	5.8	2.5	—	29.0
¾	.785	6.0	2.5	—	5.2	12.0	7.3	—	18.7
1	1.025	10.3	5.5	—	3.9	20.6	22.5	7.0	13.7
1¼	1.265	15.7	11.5	5.0	3.0	31.4	58.0	15.5	10.7
1½	1.505	22.8	28.5	8.0	2.5	45.6	109.0	38.0	8.7
2	1.935	38.6	82.0	26.0	1.8	77.2	261.0	138.0	6.3
2½	2.465	59.5	172.0	75.0	1.4	119.0	474.0	356.0	4.9
3	2.945	85.0	300.0	178.0	1.1	170.0	750.0	692.0	4.0
4	3.905	149.0	636.0	544.0	0.8	298.0	1759.0	1759.0	2.8

Fig. 20-5 Sizing tables based on velocity limitation (continued)

Copper Water Tube, Type K

Nominal Size (in.)	Actual I.D. (in.)	Flow (gpm) q	Load (wsfu) 1 * Col. A	Load (wsfu) 2 * Col. B	Friction (psi/100') p3 *	Flow (gpm) q	Load (wsfu) 1 * Col. A	Load (wsfu) 2 * Col. B	Friction (psi/100') p3 *
½	.527	2.7	.75	—	8.5	5.4	2.3	—	31.0
¾	.745	5.5	2.3	—	5.6	11.0	6.3	—	20.2
1	.995	9.7	5.3	—	4.1	19.4	19.5	5.8	14.4
1¼	1.245	15.2	10.8	5.0	3.1	30.4	54.0	14.0	11.1
1½	1.481	21.5	25.0	7.8	2.6	43.0	98.0	34.0	9.2
2	1.959	37.6	78.0	24.0	1.8	75.2	251.0	130.0	6.5
2½	2.435	58.2	166.0	69.0	1.4	116.4	460.0	340.0	5.2
3	2.907	82.8	289.0	161.0	1.2	165.6	725.0	663.0	4.2
4	3.857	146.0	609.0	528.0	0.8	292.0	1705.0	1705.0	3.0

Galvanized Iron and Steel Pipe, Standard Pipe Size

½	.622	3.8	1.5	—	8.2	7.6	3.7	—	31.0
¾	.824	6.7	3.0	—	6.0	13.4	8.4	—	22.5
1	1.049	10.8	6.1	—	4.6	21.6	25.3	7.7	17.2
1¼	1.380	18.6	17.5	6.0	3.4	37.2	77.3	23.7	12.8
1½	1.610	25.4	37.0	9.3	2.9	50.8	132.3	52.0	10.8
2	2.067	41.8	93.0	29.8	2.2	83.6	293.0	171.6	8.4
2½	2.469	59.8	174.0	75.6	1.8	119.6	477.0	361.0	6.8
3	3.068	92.0	335.0	209.0	1.4	184.0	842.0	806.0	5.4
4	4.026	158.6	688.0	615.0	1.1	317.2	1980.0	1930.0	4.1

Schedule 40 Plastic Pipe, (PE, PVC & ABS)

½	.622	3.8	1.5	—	6.8	7.6	3.7	—	24.2
¾	.824	6.7	3.0	—	5.1	13.4	8.4	—	18.0
1	1.049	10.8	6.1	—	3.7	21.6	25.3	7.7	13.2
1¼	1.380	18.6	17.5	6.0	2.8	37.2	77.3	23.7	9.6
1½	1.610	25.4	37.0	9.3	2.3	50.8	132.3	52.0	8.2
2	2.067	41.8	93.0	29.8	1.7	83.6	293.0	171.6	6.1
2½	2.469	59.8	174.0	75.6	1.4	119.6	477.0	361.0	4.8
3	3.068	92.0	335.0	209.0	1.1	184.0	842.0	806.0	3.8
4	4.026	158.6	688.0	615.0	0.8	317.2	1930.0	1930.0	2.8

*1 Col. A applies to piping which does not supply flush valves.

*2 Col. B applies to piping which supplies flush valves.

*3 Friction loss, p, corresponding to flow rate, q, for piping having fairly-smooth surface condition after extending service, applying the formula:

$$q = 4.57 (p) \quad (d)$$

Fig. 20-5 Sizing tables based on velocity limitation

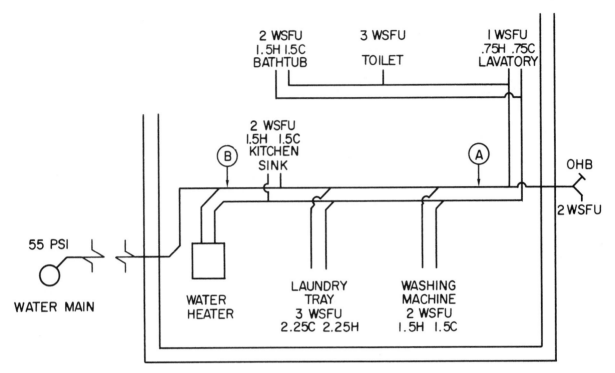

Fig. 20-6

Step 2. The schematic drawing is shown in figure 20-6.

Step 3. According to the load values assigned to fixtures, figure 20-3:

    a. Bathtub has 2 wsfu; 1.5 cold, 1.5 hot.

    b. Lavatory has 1 wsfu; .75 cold, .75 hot.

    c. Toilet has 3 wsfu; 2 cold only.

    d. Outside hose bib is not on the chart. Since it may run continuously, it should be rated at 2 wsfu.

    e. Washing machine with an 8-pound capacity has 2 wsfu; 1.5 cold, 1.5 hot.

    f. Laundry tray has 3 wsfu; 2.25 cold, 2.25 hot.

    g. Kitchen sink has 2 wsfu; 1.5 cold, 1.5 hot.

Step 4. Referring to figure 20-4, the pipe sizes from the tees to the fixtures will be:

    a. Bathtub — 1/2"

    b. Lavatory — 3/8"

    c. Toilet — 3/8"

    d. Hose bib — 1/2"

    e. Washing machine — 1/2"

    f. Laundry tray — 1/2"

    g. Kitchen sink — 1/2"

Step 5. By referring to the velocity limitation tables in figure 20-5, the rest of the building can be sized. There is a separate table for each kind of pipe. Assume that L-type copper tube is used within the house and that a K-type copper tube is used for the underground service pipe.

The tables are divided into ten columns. From left to right:

- Column 1 lists the pipe size required.

- Column 2 lists the inside pipe diameter.

- Column 3 lists the flow in gallons per minute of that size.

- Column 4 lists the load (wsfu) for tank-type toilets.

- Column 5 lists the load (wsfu) for flush valve-type toilets.

- Column 6 lists the amount of pressure which is lost in 100 feet.

- Columns 7, 8, 9, and 10 are the same as columns 3, 4, 5, and 6. Columns 7, 8, 9, and 10, however, are for velocities of 8 feet per second. Columns 3, 4, 5, and 6 are for velocities of 4 feet per second.

RESULTS: In step 4, the branches were sized from the fixtures to the tees which serve them. From the bathroom group to point A on figure 20-6 there are 7.25 wsfu on the cold side. According to the tables under the 8 fps and flush tank toilets column, 1/2-inch tubing will handle 2.5 wsfu and 3/4-inch tubing will handle 7.3 wsfu.

Step 1. From point A to the bathroom group, there should be 3/4-inch tubing on the cold side. Therefore, there are 2.25 wsfu from the tee at point A to the tee in the bathroom group on the hot side.

Step 2. From point A to the bathroom group, there should be 1/2-inch tubing on the hot side. From point B to point A the pipe must carry the kitchen sink, the washing machine, the laundry tray, plus the bathroom group and hose bib. On the cold side there are 12.50 wsfu. On the hot side there are 7.5 wsfu.

Step 3. From point B to point A on the cold side, there should be 3/4-inch tubing. 12.5 is closer to 7.3 than it is to 22.5.

Step 4. From point B to point A, there should be 3/4-inch tubing on the hot side. From point B to the street main, the cold line must carry all of the load for the building, both hot and cold. When this occurs, the total figure per fixture must be used rather than adding the individual hot and cold figures. There are 25 total wsfu for the building. Types L and K tubing will carry 22.5 and 19.5 in the 1-inch size respectfully.

Step 5. Therefore, the water service and the main to point B will be in the 1-inch size using this method.

REVIEW QUESTIONS

Multiple Choice

Select the best answer for each question.

1. The loss of pressure from one point on a line of pipe to a point downstream is known as

   a. overhead.

   b. friction loss.

   c. corrosion effect.

   d. red line shift.

2. As the size of the pipe increases with the same available pressure,

   a. the water flows more rapidly.

   b. the water slows down.

   c. the cost of the system goes down.

   d. none of the above.

3. Hardness of water is another name for

   a. mineral content.

   b. salt balance.

   c. flowing qualities.

   d. silt and sediment content.

4. The pH factor of water refers to the

   a. mineral content.

   b. acid balance.

   c. flowing qualities.

   d. silt and sediment content.

5. Under normal conditions, the velocity of the water should be _____ feet per second.

   a. 4
   c. 12
   b. 8
   d. 15

6. Which of the following pipe materials will carry the greatest load in the 1-inch size?

   a. Copper water tube, type L
   b. Copper water tube, type K
   c. Plastic pipe, schedule #40
   d. Copper and brass pipe, standard size

Questions

1. What is the supply pipe size for the following fixtures?

   a. Bathtub
   b. Tank-type toilet
   c. Lavatory
   d. Kitchen sink
   e. Hose bib

2. What is the private water supply load value for the following fixtures?

   a. Bathtub
   b. Tank-type toilet
   c. Lavatory
   d. Kitchen sink
   e. Hose bib

3. What is the wsfu for the following fixtures on the cold side only?

   a. Bathtub (private)
   b. Lavatory (private)
   c. Shower head (public)

Project

Size the pipe (K copper tube) for the home in figure 20-7. The city water is at 60 psi. There is no record of excessive corrosion.

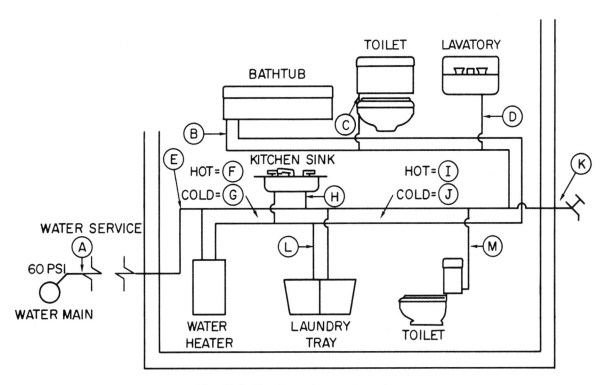

Fig. 20-7 Size the water supply system

Drainage Systems

UNIT 21 FIXTURE UNITS

OBJECTIVES

After studying this unit, the student will be able to:

- explain what a fixture unit is.

- describe how fixture units apply to various plumbing fixtures.

FIXTURE UNITS

To determine the amount of discharge a drainage system will accommodate, the Bureau of Standards developed the fixture-unit method after intensive research. It is based upon the discharge of a lavatory, which is 7 1/2 gallons per minute or one cubic foot per minute. This is called a *fixture unit*. The fixture unit is the basis for planing and designing plumbing systems. All other fixtures are rated according to their discharge as compared to one fixture unit.

Table 21-1 shows the number of fixture units for a variety of fixtures and the sizes of traps and waste or soil pipes required. For fixtures which flow continuously, such as air-conditioning equipment and sewage ejectors, allow 2 fixture units for each gallon per minute of flow.

The *maximum fixture-unit load* is the greatest number of fixture units that may be connected to a given size of a house drain, horizontal branch, vertical soil, or waste stack. Tables 21-2 and 21-3, page 88, give maximum fixture-unit loads.

Type of Fixture or Group of Fixtures	Drainage Fixture-Unit Value (d.f.u.)
Automatic clothes washer (2" standpipe)	3
Bathroom group consisting of a water closet, lavatory and	
bathtub or shower stall:	
Flushometer valve closet	8
Tank-type closet	6
Bathtub[1] (with or without overhead shower)	2
Bidet	1
Clinic Sink	6
Combination sink-and-tray with food waste grinder	4
Combination sink-and-tray with one 1-1/2" trap	2
Combination sink-and-tray with separate 1-1/2" traps	3
Dental unit or cuspidor	1
Dental lavatory	1
Drinking fountain	1/2
Dishwasher, domestic	2
Floor drains with 2" waste	3
Kitchen sink, domestic, with one 1-1/2" trap	2
Kitchen sink, domestic, with food waste grinder	2
kitchen sink, domestic, with food waste grinder and	
dishwasher 1-1/2" trap	3
Kitchen sink, domestic, with dishwasher 1-1/2" trap	3
Lavatory with 1-1/4" waste	1
Laundry tray (1 or 2 compartments)	2
Shower stall, domestic	2
Showers (group) per head	2
Sinks:	
Surgeon's	3
Flushing rim (with valve)	6
Service (trap standard)	3
Service (P trap)	2
Pot, scullery, etc.	4
Urinal, pedestal, syphon jet blowout	6
Urinal, wall lip	4
Urinal, stall, washout	4
Urinal trough (each 6-ft. section)	2
Wash sink (circular or multiple) each set of faucets	2
Water closet, tank-operated	4
Water closet, valve-operated	6
Fixtures not listed above:	
Trap Size 1-1/4" or less	1
Trap Size 1-1/2"	2
Trap Size 2"	3
Trap Size 2-1/2"	4
Trap Size 3"	5
Trap Size 4"	6

[1] A shower head over a bathtub does not increase the fixture-unit value

Table 21-1 Drainage fixture-unit values for various plumbing fixtures

Maximum Number of Fixture Units that may be Connected to:

Diameter of Pipe	Any Horizontal[1] Branch Interval	One Stack of Three Branch Intervals or Less	Stacks with More Than Three Branch Intervals	
			Total for Stack	Total at One Branch Interval
Inches				
1½	3	4	8	2
2	6	10	24	6
2½	12	20	42	9
3	20[2]	48[2]	72[2]	20[2]
4	160	240	500	90
5	360	540	1,100	200
6	620	960	1,900	350
8	1,400	2,200	3,600	600
10	2,500	3,800	5,600	1,000
12	3,900	6,000	8,400	1,500
15	7,000			

[1] Does not include branches of the building drain.
[2] Not more than 2 water closets or bathroom groups within each branch interval nor more than 6 water closets or bathroom groups on the stack.

Table 21-2 Horizontal fixture branches and stacks

Diameter of Pipe	Maximum Number of Fixture Units That May Be Connected to Any Portion of the Building Drain or the Building Sewer Including Branches of the Building Drain.[1]			
	Fall Per Foot			
	1/16-Inch	1/8-Inch	1/4-Inch	1/2-Inch
Inches				
2			21	26
2½			24	31
3		36*	42*	50*
4		180	216	250
5		390	480	575
6		700	840	1,000
8	1,400	1,600	1,920	2,300
10	2,500	2,900	3,500	4,200
12	2,900	4,600	5,600	6,700
15	7,000	8,300	10,000	12,000

*Not over two water closets or two bathroom groups.
1. On site sewers that serve more than one building may be sized according to the current standards and specifications of the Administrative Authority for public sewers.

Table 21-3 Building drains and sewers

REVIEW QUESTIONS

Multiple Choice

Select the best answer for each question.

1. Who developed the fixture-unit method?

   a. National Plumbing Code
   b. The Bureau of Standards
   c. The American Society of Plumbing Engineers
   d. NASA

2. What is the fixture unit based upon?

   a. The fixture height above street level
   b. The volume of the fixture-holding capacity
   c. The discharge of a lavatory
   d. Water viscosity

3. How many fixture units does a home have if there is one bathroom with a tank-type water closet, a lavatory, a bathtub with shower, a powder room with a tank-type water closet and a lavatory, a kitchen with a sink and dishwasher, and a laundry room with a double compartment tray and an automatic clotheswasher with a standpipe?

   a. 15   c. 21
   b. 19   d. 32

4. How many fixture units can a 3-inch stack handle with three or less branch intervals?

   a. 9   c. 20
   b. 15   d. 48

5. How many fixture units can a 3-inch building drain handle if it has a 1/4 inch per foot fall?

   a. 25   c. 42
   b. 36   d. 48

6. How many water closets can be installed to a 3-inch stack on the first floor?

   a. 1   c. 6
   b. 2   d. 8

7. How many fixture units can a 4-inch stack in a two-story home handle?

   a. 48   c. 120
   b. 52   d. 240

8. How many fixture units can a 4-inch house drain at a 1/4 inch per foot fall handle?

   a. 180   c. 250
   b. 216   d. 430

9. What is the fixture unit rating of a 3-inch trap?

    a. 2   c. 6
    b. 5   d. 7

10. How many gallons of water should a lavatory discharge in the space of one minute?

    a. 5 gallons   c. 10 gallons
    b. 7.5 gallons   d. 15 gallons

UNIT 22 SOIL STACKS
AND STACK INLET FITTINGS

OBJECTIVES

After studying this unit, the student will be able to:

• explain how stacks are installed and how water flows in them.

• determine the capacity of fittings in gallons per minute.

STACKS

A *stack* is a vertical drainpipe. A *soil stack* is a vertical drainpipe which receives the discharge from toilets. Other fixtures may also drain into it.

Soil stacks extend full size throughout their entire length. They extend above a roof not less than 6 inches and not more than 24 inches without support. A stack running within 10 feet of any door, window, or air shaft cannot end less than 2 feet above the door, window, or air shaft.

Soil stacks are constructed of cast-iron soil pipe and fittings with hubless coupling joints, compression joints, lead and oakum joints, galvanized wrought-iron pipe with recessed drainage fittings, or copper tubing with sweated fittings.

All offsets on vent stacks are made with 45-degree bends to prevent the pipe from clogging with accumulated rust scale. The stacks enter the main drain at an angle of 45 degrees. If the stack is not directly over the drain, a long sweep bend with a clean-out may be used.

Stacks are supported at the base by piers or hangers. In large buildings, additional supports are placed at each floor by means of beam clamps, figure 22-1.

In small buildings, soil stacks are placed inside partitions or in a pipe chase. The wall must have at least a 7-inch space inside to accommodate a 4-inch pipe hub, figure 22-2. Pipe shafts are also used in very large

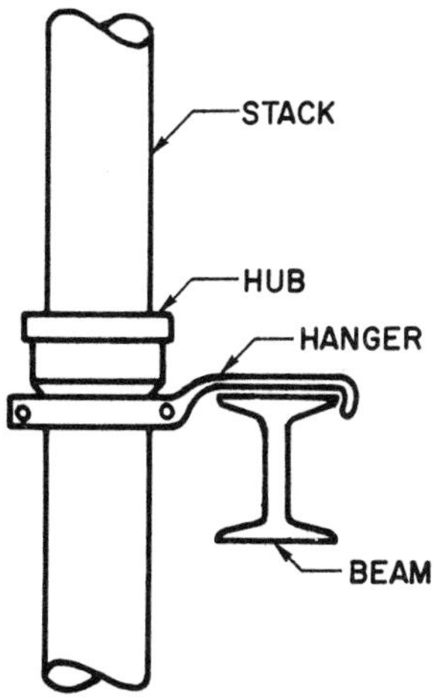

![Fig. 22-1 Supporting a stack]
STACK
HUB
HANGER
BEAM

Fig. 22-1 Supporting a stack

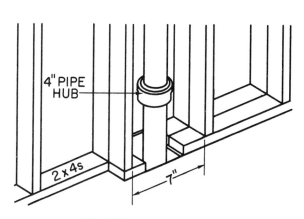
4" PIPE HUB
2 x 4s
7"

Fig. 22-2

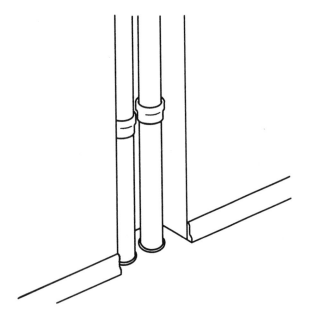

Fig. 22-3

Size in In.	Fixture Units	Toilets	Lengths in Feet
1 1/2	4		50
2	10		75
2 1/2	20		100
3	48		150
4	240	33	300
5	540	80	500
6	960	120	unlimited
8	2200	225	unlimited
10	3800	400	unlimited

Fig. 22-4 Stack sizes

buildings to provide easy access if repairs are needed, figure 22-3.

Roof flanges are placed around stacks where they pass through the roof to prevent leakage. In cold climates, small stacks (less than 3 inches) must be increased in size where they pass through roofs to prevent them from being closed by frost. Stacks should never be placed outside of buildings. Such stacks are unsightly and may freeze closed with frost.

The size stack required for various combinations of fixture units and toilets is shown in figure 22-4.

STACK INLET FITTINGS

Sanitary tees are placed in stacks for branch connections because of the limited space between floors and ceilings. Whenever possible, the combination wye and eighth bend is used. This offers better drainage at a point just below the inlet, figure 22-5.

Stacks do not run completely full of water. Therefore, the capacity and velocity depend upon other factors besides the pipe size. The stack capacity is affected by the

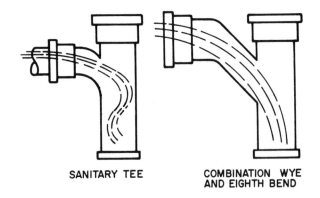

SANITARY TEE   COMBINATION WYE
                      AND EIGHTH BEND

Fig. 22-5

inlet fitting. Water entering the sanitary tee splashes directly against the opposite side of the stack. This fills the stack directly below the branch and retards the flow. Water entering the combination wye and eighth bend strikes the opposite side of the stack at an angle and does not retard the flow. Therefore, a 3-inch combination wye and eighth bend has more capacity than a 4-inch sanitary tee.

The formula for determining capacity of fittings is $C = KD^2$. C is the capacity in gallons per minute. D is the diameter. K is a factor which varies for each type of fitting. K for a 45-degree fitting is 22.5; K for a sanitary tee is 11.25.

Therefore, the capacity of a 3-inch combination wye and eighth bend is 22.5 X 3 X 3 which equals 222.5 gallons per minute. The capacity of a 4-inch sanitary tee is 11.25 X 4 X 4 which equals 180 gallons per minute.

Water flowing down stacks is delayed by friction and air. It also forms slugs and diaphragms around the solids which slow the velocity, so the height of stacks need not be limited on that account. However, the stack should allow room for air movement and for relief vents at the base.

The base fitting should be a 45-degree sweep or a long sweep bend. These cause less friction. Back pressure may result if this point is neglected.

REVIEW QUESTIONS

Multiple Choice

Select the best answer for each question.

1. Main stacks must extend

   a. to the farthest water closet.
   b. 25 feet from the nearest vent.
   c. full size throughout their length.
   d. to the nearest water closet.

2. A stack which terminates within 10 feet of a window

   a. must have a roof flange.
   b. must have at least 3 wire supports with turnbuckles.
   c. must open 2 feet above that window.
   d. none of the above.

3. All offsets in a vent stack must

   a. have a 60-degree angle.
   b. consist of a single fitting.
   c. be washed out by at least one fixture.
   d. be no sharper than 45 degrees.

4. The purpose of a pipe chase is to

   a. find the material to put on the truck.
   b. provide easy access to the pipe.
   c. swab out the internal surfaces of the pipe.
   d. thread pipe.

5. A 4-inch stack will accommodate how many fixture units?

   a. 240 fixture units
   b. 320 fixture units
   c. 540 fixture units
   d. None of the above

6. How many gallons per minute will be drained by a 5-inch combination wye and eighth bend?

   a. 118.25 gallons per minute
   b. 222.5 gallons per minute
   c. 281.25 gallons per minute
   d. 562.5 gallons per minute

7. How many gallons per minute will be drained by a 6-inch sanitary tee?

   a. 180 gallons per minute
   b. 222 gallons per minute
   c. 405 gallons per minute
   d. 810 gallons per minute

8. The purpose of a roof flange is to

   a. support the top of the stack.
   b. keep rain from entering the building through the stack hole.
   c. insulate the stack from the building structure.
   d. keep rain out of the stack.

9. An offset in a soil stack can be much sharper than one in a vent stack because

   a. scale does not form in the presence of water.
   b. air currents do not make sharp bends.
   c. the offset on a soil stack is washed out.
   d. none of the above.

10. What special precautions should be taken to guard against the closing of stacks in very cold climates?

   a. The stack should be run in a heated pipe chase.
   b. Stacks must be at least 3 inches where they pass through the roof.
   c. Stacks should extend no more than 6 inches through the roof.
   d. The valve should be locked open.

UNIT 23 WASTE AND VENT STACKS

OBJECTIVES

After studying this unit, the student will be able to:

- describe the different types of waste stacks and their use.
- describe the purpose and use of vent stacks.

WASTE STACKS

A *waste stack* is a vertical drainpipe which receives the discharge of small fixtures. *Small fixtures* can be any plumbing fixture except toilets. Stacks may be constructed with the same pipe and fittings as other drainpipes and are subject to the same regulations as other stacks. The number of fixtures and fixture units permitted on waste stacks may be found in the local code.

BRANCH SOIL AND WASTE PIPES

A *branch waste pipe* is a horizontal or vertical pipe which receives the discharge of small fixtures and conveys it to the stack.

If such a pipe receives the discharge of toilets and urinals, it is called a *branch soil pipe*. The size of this pipe is determined by the fixture-unit method.

A branch for a single fixture is shown in figure 23-1. The distance from the trap to the vent is called the *trap arm*. The maximum length of the trap arm can be determined by using the chart in figure 23-2, page 96.

Branches should have a pitch of 1/4 inch or more per foot and enter the stack or main drain at a 45-degree angle. A cleanout is placed on the end of the branch. If more than one fixture is on the line, it must be vented.

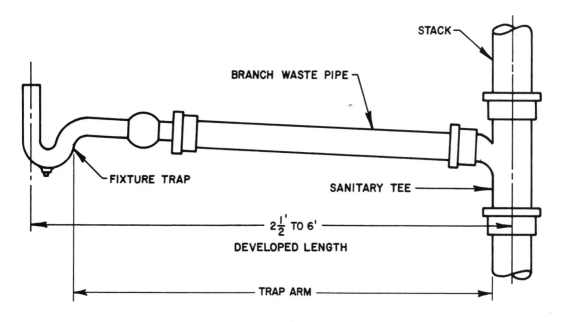

Fig. 23-1 Branch waste pipe

Size of Fixture Drain Inches	Distance — Trap to Vent
1¼	2 ft. 6 in.
1½	3 ft. 6 in.
2	5 ft.
3	6 ft.
4	10 ft.

Fig. 23-2 Maximum length of trap arm

The length of the branch waste pipe is limited to 2 1/2 to 6 feet developed length. The *developed length* is the distance measured along the centerline of the pipe, from the trap to the vent. Some local codes may vary in their allowable developed length.

BRANCH CONNECTIONS TO STACKS

A sewage system is designed to remove sewage from a building quickly and without breaking it up. Branch soil and waste pipes connect to the main drain at an angle of 45 degrees. This directs the sewage in the direction of the flow in the main drain. Single or double wyes or combination wye and eighth bends may be used for this purpose. Fittings

in the main drain used as branches should always pitch up to allow the branch at least 1/4 inch per foot pitch, figure 23-3.

On stacks, sanitary tees and short-turn tees may be used for branches. These are not the most effective but are permitted because of the limited space between floors and ceilings. Short-turn tees may retard the flow directly under the branch in stacks and cause slugs to form.

Branch soil or waste pipes may be made of cast iron, brass, wrought iron, galvanized steel, copper, ABS and PVC plastic, or lead. If lead is used, it is laid on a board to prevent sagging.

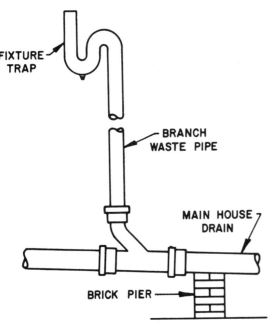

Fig. 23-3

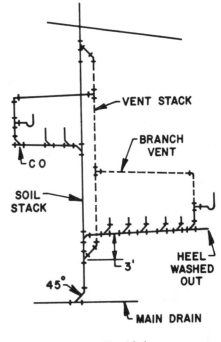

Fig. 23-4

VENT STACKS

A vent pipe has two purposes. First, it provides a continuous change of fresh air within a drainage system. Second, it prevents siphoning and back pressure. *Back pressure* is caused by descending slugs of water in a stack which may break the trap seal. By providing fresh air to the drainage system, the accumulation of gases is prevented. Otherwise, these gases are absorbed by the moisture clinging to the top of horizontal pipe and speed up corrosion.

A vent stack is connected to the soil or waste stack with a wye fitting at or below the lowest branch, figure 23-4. Connecting the bottom of the vent stack into the soil stack in this way permits rust scale to be washed away. It also relieves back pressure at this point.

A vent stack extends at least 6 inches above the flood rim of the highest fixture where it is reconnected to the stack vent. If the vent is more than 25 feet away, it must pass through the roof separately. In cold climates, small stacks must be increased to at least 3 inches before passing through the roof to prevent freezing.

The size of vent stacks and branches, permitted fixture units, and the developed length in feet is shown in figure 23-5.

Size of soil or waste stack	Fixture Units Con-nected	Diameter of Vent Required (Inches)								
		1-1/4	1-1/2	2	2-1/2	3	4	5	6	8
		Maximum Length of Vent (Feet)								
Inches										
1½	8	50	150							
1½	10	30	100							
2	12	30	75	200						
2	20	26	50	150						
2½	42		30	100	300					
3	10		30	100	100	600				
3	30			60	200	500				
3	60			50	80	400				
4	100			35	100	260	1000			
4	200			30	90	250	900			
4	500			20	70	180	700			
5	200				35	80	350	1000		
5	500				30	70	300	900		
5	1100				20	50	200	700		
6	350				25	50	200	400	1300	
6	620				15	30	125	300	1100	
6	960					24	100	250	1000	
6	1900					20	70	200	700	
8	600						50	150	500	1300
8	1400						40	100	400	1200
8	2200						30	80	350	1100
8	3600						25	60	250	800
10	1000							75	125	1000
10	2500							50	100	500
10	3800							30	80	350
10	5600							25	60	250

Fig. 23-5 Size and length of vents

REVIEW QUESTIONS

Multiple Choice

Select the best answer for each question.

1. Why is a 45-degree elbow used to make an offset in a waste stack?
   a. To prevent rust from falling into the main drain
   b. So that a distance of 6 inches can be maintained between stacks
   c. To prevent back pressure
   d. To direct the flow quickly and without breaking it up

2. Describe the fitting used to connect a 2-inch galvanized branch to a 3-inch galvanized stack.
   a. A 2″ x 3″ tapped tee   c. A 3″ x 2″ sanitary tee
   b. A 2″ x 3″ sanitary tee   d. A 3″ x 2″ bullhead tee

3. How will lack of venting affect the metal pipe in a drainage system?
   a. Gases will combine with moisture to cause corrosion.
   b. Odors will collect in the living spaces.
   c. Condensation will form on all pipe surfaces.
   d. There would be no effect.

4. Describe the fitting in the lower end of a 5-inch soil stack to which a 3-inch vent stack is to be attached.
   a. A 5″ x 3″ sanitary tee
   b. A 5″ x 3″ combination wye and eighth bend
   c. A 5″ x 3″ upright wye
   d. None of the above

5. A branch _____ pipe receives the discharge from small fixtures other than a toilet or urinal.
   a. waste   c. vent
   b. soil   d. water

6. A branch _____ pipe receives the discharge of toilets or urinals.
   a. waste   c. vent
   b. soil   d. water

7. What is the smallest vent that could be used on a 2-inch waste stack to which 12 fixture units are drained?
   a. 1-inch vent   c. 2-inch vent
   b. 1 1/4-inch vent   d. 3-inch vent

8. How long could the vent be in question #7?
   a. 5 feet   c. 20 feet
   b. 10 feet   d. 30 feet

9. How far away from its vent could a 1 1/2-inch trap be placed?
   a. 2'-6″   c. 5'-0″
   b. 3'-6″   d. 8'-0″

10. The developed length of a drain is measured

    a. between the trap outlet and the stack across the floor.
    b. from the trap to the vent following the centerline of the pipe.
    c. on the blueprint with a scaling rule.
    d. by using appropriate tables.

UNIT 24 LOSS OF TRAP SEALS

OBJECTIVES

After studying this unit, the student will be able to:

- describe siphonage and its effect on various types of traps.
- describe back pressure and how to prevent it.
- discuss capillary attraction and evaporation.

SIPHONAGE

A siphoned trap is a health problem to the occupants of any building. It is important to understand the siphonage of trap seals so that it may be prevented.

A siphon is a bent tube with arms of unequal length. If the short end is inserted in water and air is exhausted from the tube, atmospheric pressure will force the water up the short arm and over the crown. The water will continue to run until the receptacle is empty, figure 24-1.

A *plumbing trap* is similar to a siphon. The trap may be siphoned when connected to a waste pipe. Water will stay in the trap when the atmospheric pressure, about 15 psi, is the same on both the inlet and the outlet, figure 24-2.

When water descends in a waste stack, it picks up considerable velocity. This causes a reduction in air pressure at the top of the stack. If this air cannot be replaced, the trap will be siphoned, figure 24-3. The causes may be improper design or size of the stack; an obstruction by rust, scale, or hoar frost; or the omission of a vent pipe.

Water and sewage descending a soil stack tend to form slugs. These slugs cause an air pressure greater than the atmospheric pressure, called a *plenum*, at the bottom, and a

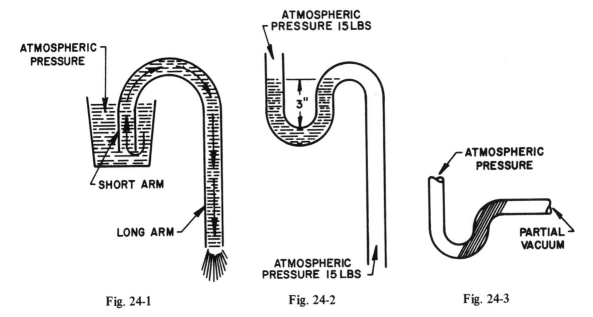

ATMOSPHERIC PRESSURE

SHORT ARM

LONG ARM

Fig. 24-1

ATMOSPHERIC PRESSURE 15LBS

3"

ATMOSPHERIC PRESSURE 15 LBS

Fig. 24-2

ATMOSPHERIC PRESSURE

PARTIAL VACUUM

Fig. 24-3

lower air pressure or *partial vacuum* at the top of the stack. This condition expels the water from the lower traps and siphons the trap seals on the upper floors. Both conditions admit drainage air to rooms, thereby presenting a health hazard.

A properly installed vent pipe furnishes air at the top and relieves air at the bottom. This equalizes the atmospheric pressure within the stack. Notice the slug of water in figure 24-4 as it passes the upper trap outlet. The slug lowers the air pressure in the branch, and atmospheric pressure drives out the trap seal. At the lower trap, it creates an excess pressure which drives the water out of the trap.

When water passes the trap outlet, figure 24-5, air pressure from the vent pipe relieves the partial vacuum. The seal remains in the trap.

The average depth of a seal in an S trap is about 3 inches. One foot of water exerts a pressure of .434 psi; 3 inches will exert 1/4 of .434 or .108 psi of pressure. This shows what a small amount of pressure a trap seal can withstand. On waste pipes which drop

vertically for some distance below a trap, the discharge of a fixture may self-siphon its own trap.

Siphonage of traps can be prevented by checking the sizes and lengths of the branch waste pipe and installing the correct vent pipes. The best trap and waste pipe connection is the 1/2 S trap connected to the continuous vent, as shown in figure 24-6. The vent pipe supplies all the air needed, and siphonage is prevented. However, the branch should be kept short so that the top of the branch outlet is never below the dip of the trap, figure 24-7, page 102.

The vent pipe should never be attached to the crown of the trap because grease and other matter will soon clog the vent pipe, figure 24-8, page 102.

Oval-bottomed fixtures discharge water more quickly than flat ones. Therefore, the former are more likely to be siphoned. In the latter, the seal is retained by the trickle of water to the trap after the main discharge has passed. The *drum trap*, in the form of the Philadelphia regulation bath trap, cannot

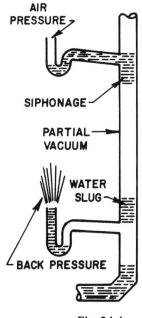

Fig. 24-4

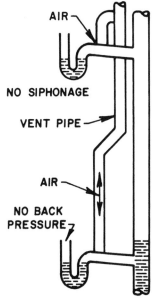

Fig. 24-5

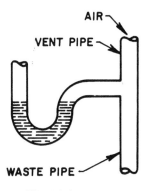

Fig. 24-6

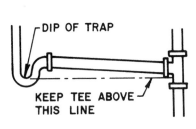
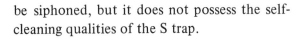

Fig. 24-7

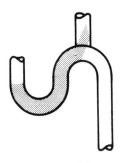

Fig. 24-8

be siphoned, but it does not possess the self-cleaning qualities of the S trap.

BACK PRESSURE

The opposite action of siphonage is *back pressure*. Back pressure usually occurs at the base of stacks when water slugs, which form in soil or waste stacks, push air before them. This has a tendency to increase the air pressure at the base of the stack, figure 24-9. The air pressure will increase if the horizontal drain is partially filled from other fixtures, or if it receives rainwater during a storm. This may not be as harmful as siphonage since, in most cases, the water will run back into the trap.

Back pressure is prevented by installing a vent pipe or a relief pipe in the stack below the lowest fixture. The main vent or relief vent is connected to the soil stack with a wye connection. This prevents the collection of rust scale at that point. This pipe is usually reconnected to the soil stack above the highest fixture. Back pressure is then eliminated because the air can escape through the relief vent.

The fitting used at the base of the stack affects the pressure at that point. A tee is the poorest fitting. A combination wye and eighth bend, which is most often used, is better. A long-sweep bend is best.

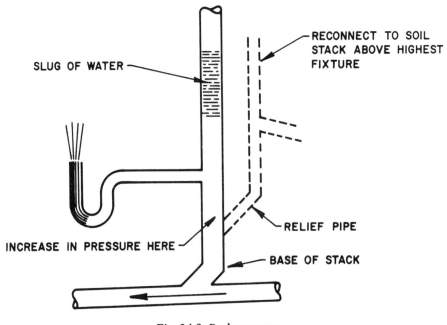

Fig. 24-9 Back pressure

CAPILLARY ATTRACTION

Capillary attraction is the power that small tubes, string, and other porous materials have to lift water above their own level. For example, sap rises in trees through the small tubes in the wood.

If a piece of string or lint catches in a trap, and one end hangs over into an outlet, it may act as a wick and carry the water over by drops until the seal is broken, figure 24-10. This would allow sewer gas to enter the room. The inside of traps should be smooth to prevent material from catching.

EVAPORATION

Evaporation is the change of water to vapor. Evaporation is more rapid at high temperatures. Air currents increase evaporation. The trap seal of an unused plumbing fixture may evaporate and admit sewer air to the house. There is no danger when a fixture is used frequently. A vented trap is more likely to evaporate than an unvented one.

Trap seals may be lost by evaporation if unused for long periods. The length of time depends upon the amount of humidity contained in the air, the temperature, whether the air is still or moving, and the surface area of the water.

The traps affected most are rain conductors in summer and floor drains in winter. Long periods of drought may cause rain conductor traps to lose their seals. For this reason, deep seal traps are required on all rain conductors. The seal of a 4-inch trap should be 6 inches deep and maintain a seal for about 3 months.

When floor drains are installed inside buildings, provision must be made to supply them with water by placing a faucet nearby, figure 24-11. Yard and area drains are also subject to evaporation, the result being objectionable odors. However, since they are outside, they are not dangerous. The seal may be replaced by adding water.

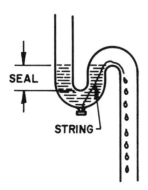

Fig. 24-10 Capillary attraction

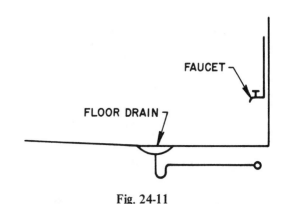

Fig. 24-11

REVIEW QUESTIONS

Multiple Choice

Select the best answer for each question.

1. A siphon is
    a. a tube shaped like a 1/2 S trap.
    b. a bent tube with arms of unequal length.
    c. a tube with a pressure of 15 psi on one end.
    d. a rubber hose.

2. The discharge from a fixture may form a _____ as it descends in the stack.

   a. slug
   b. trap

   c. gradient
   d. vent

3. A pressure that is higher than atmospheric pressure in a stack is called

   a. a vacuum.
   b. a siphon.

   c. a plenum.
   d. a gradient.

4. The average depth of seal in a fixture trap is about

   a. 6 inches.
   b. 5 inches.

   c. 4 inches.
   d. 3 inches.

5. A vent pipe should never be attached to the crown of a trap because

   a. the trap will rust out.
   b. it promotes siphonage of the trap.
   c. it is then impossible to remove the trap for service.
   d. It will soon clog with grease and other matter.

6. The seal of a deep seal trap is most nearly

   a. 2 1/2 inches.
   b. 3 inches.

   c. 4 5/8 inches.
   d. 6 inches.

7. The loss of trap seal is prevented for floor drains inside buildings by

   a. proper venting.
   b. a faucet placed nearby.
   c. a cleanout placed within 18 inches of the trap.
   d. occasional rainfall.

8. What conditions must exist in a drainage system so that water remains in the traps?

   a. Equal pressure on both sides of the trap
   b. Proper fall in all drain lines
   c. High humidity and low temperatures
   d. Temperatures below freezing

9. When water descends a stack, air is drawn in behind it. Where does this air come from?

   a. From the yard vent
   b. From the traps

   c. From the stack vent
   d. Below the slug

10. If the air in a stack is not replaced quickly, what is likely to happen?

    a. The slug of water will stop.
    b. Air will be drawn through the traps.
    c. Trap seals in the lower floors will be blown out.
    d. The stack will collapse.

11. What does a vent pipe supply to stacks?
     a. Structural soundness   c. A passageway for waste
     b. A passageway for air to travel d. A plenum

12. Pressure below atmospheric pressure is known as
     a. siphonage.   c. a plenum.
     b. vent effect.   d. a vacuum.

13. What is the top of a vent pipe reconnected to?
     a. An S trap   c. The stack vent
     b. The roof scupper   d. The base of the soil stack

14. What causes back pressure in a drainage system?
     a. Wind blowing over the top of the stack
     b. A descending slug of water
     c. Evaporation
     d. Loss of trap seal

15. How is the accumulation of rust and scale at the base of a vent stack prevented?
     a. By using deep-seal traps
     b. By connecting the base of the stack with a wye fitting
     c. By the occasional use of a plumber's solvent
     d. By using rubber rings

16. What happens when back pressure occurs to a trap?
     a. The water is siphoned out of the trap.
     b. The water is momentarily forced up into the fixture.
     c. Air is drawn through the trap into the drainage system.
     d. The trap swells.

17. How may back pressure be prevented?
     a. By installing a relief vent below the lowest fixture
     b. Cleaning out the stack vent
     c. Tying rain conductors into the building drain
     d. By using roof flanges

Plumbing Sketches

1. Sketch a fixture trap which is similar to a siphon.

2. Sketch the most effective trap and vent connection.

UNIT 25 CONTINUOUS VENTS AND WET VENTS

OBJECTIVES

After studying this unit, the student will be able to:

- explain the importance of continuous vents.
- describe the uses of wet vents.

CONTINUOUS AND WET VENTS

A *continuous vent* is a continuation of the vertical waste pipe up, past the fixture, and back into a branch vent, vent stack, or stack vent, figure 25-1. The use of the continuous vent, also known as the *back vent*, is probably the most effective method of venting. This vent may be constructed of lead, brass, copper, galvanized wrought iron, cast iron, or plastic pipe (where permitted by local code).

The size of the vent pipe should be no less than 1/2 the waste pipe and never less than 1 1/4 inch. The tee in the waste pipe must not be placed below the trap seal or siphonage may occur.

In back venting, the vertical pipe is close to the fixture trap. The vent pipe extends to a point above the fixture before connecting to the branch vent, or it connects to the stack above the highest fixture.

If the fixture is at some distance from the vertical waste pipe, the branch waste pipe should be extended to a point nearer the fixture, figure 25-2. If the vent is run as shown in figure 25-3, the waste backs up in the vent pipe, soap or grease is deposited, and the vent is eventually closed. The water level in the waste and vent lines assumes the hydraulic gradient level when water is released from the fixture.

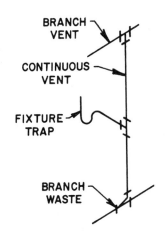

Fig. 25-1

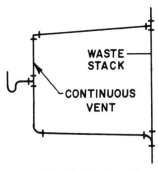

Fig. 25-2 (Correct)

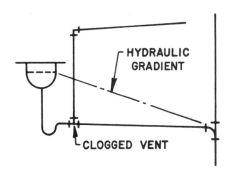

Fig. 25-3 (Incorrect)

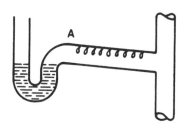

Fig. 25-4

Another reason for keeping the horizontal pipe short is that air will not circulate in a branch pipe more than 1 or 2 feet. Moisture collects on the top of the pipe when it is not ventilated. This moisture absorbs carbonic acid gases and rusts the pipe along the top.

Figure 25-4 shows how moisture clings to and destroys the pipe. A vent pipe placed at A would remove the gases and prevent destructive action on the pipe.

Fixtures on three or more floors should be connected as shown in figure 25-5. It is not necessary to vent the fixtures on the top floor since there are no fixtures above them.

THE WET VENT

A *wet vent* is a vent for one fixture which is used for a drain by a fixture above. The continuous vent and a soil stack in a small structure answer this description. However, wet vents are used only in two-story buildings, figure 25-6. The upper fixtures, which drain into the waste stack, do not require an additional vent as long as trap arms are not longer than the maximum length allowed. The lower fixtures drain into the bottom of the vent stack. They also do not require an additional vent.

In all stacks it is best to use short-turn, single or double-drainage TYs. If long-turn TYs are used where the branch is long (5 to 8 feet), the waste pipe at the junction of the stack may be below the dip of the trap, thereby causing siphonage. The pipe size depends upon the number of fixtures.

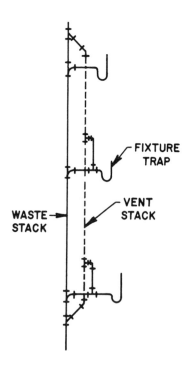

Fig. 25-5

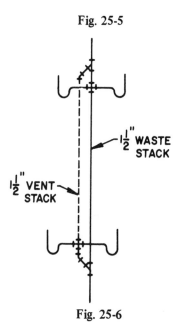

Fig. 25-6

A wet vent must be a minimum of 1 1/2 inches if one lavatory drains into it. A 2-inch wet vent may handle up to 4 fixture units. In both cases, the horizontal drain line should be 2 inches. On the lower floors of a multi-story building, a lavatory drain may be used as a vent if the drain and vent are at least 2 inches.

REVIEW QUESTIONS

Multiple Choice

Select the best answer for each question.

1. Which one of the following fittings is the best choice for connecting into the waste stack?

    a. Long-turn TY

    b. Short-turn TY

    c. Tapped tee

    d. A vented wye

2. What size vent should be used for a lavatory?

    a. 3/4-inch vent

    b. 1 1/4-inch vent

    c. 2-inch vent

    d. 2 1/2-inch vent

3. What is the purpose of a vent pipe?

    a. To provide a passage for waste

    b. To prevent corrosion and the loss of trap seals

    c. To support fixtures

    d. To prevent water hammer

4. Why is it important to keep the tee in the waste pipe as high as the dip in the trap?

    a. So the trap arm does not become a siphon

    b. To keep the trap arm within the space provided

    c. To prevent the trap from drying out

    d. To keep the pipe between the ceiling and the floor

5. How high should the branch vent be placed?

    a. Below the ceiling

    b. At least 6 inches below the flood rim of the fixtures it serves

    c. Even with the hydraulic gradient

    d. Above the fixtures it serves

6. To what is the branch vent in figure 25-1 extended?

    a. To the fixture

    b. To the soil stack

    c. To the vent stack

    d. None of the above

7. What could be a problem with the connections in figure 25-3?

    a. The trap could leak.

    b. Carbonic acid could be formed.

    c. The trap could be siphoned.

    d. The vent connection could become clogged.

8. At a fall of 1/4 inch per foot, in how many feet will the top of the opening in a 2-inch pipe be even with the bottom?

    a. 12 feet

    b. 8 feet

    c. 4 feet

    d. 2 1/2 feet

9. Why should the branch to a single fixture be short?

   a. To keep material costs down
   b. So that the stack inlet is not below the dip of the trap
   c. To avoid weakening the building structure
   d. To minimize rust

10. How many fixture units will a 2-inch wet vent handle?

    a. 1   c. 8
    b. 4   d. 10

Plumbing Sketch

Make a simple line drawing showing how to back vent the lavatory in figure 25-7. Indicate the size of the vent.

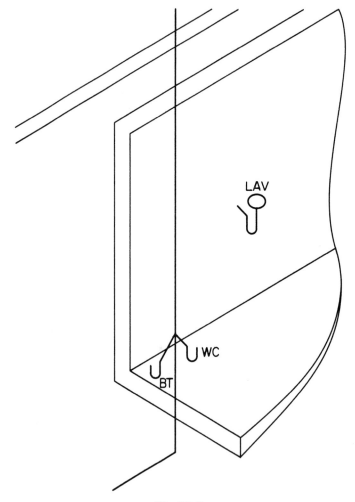

Fig. 25-7

UNIT 26 LOOP AND CIRCUIT VENTS

OBJECTIVE

After studying this unit, the student will be able to:

- describe loop and circuit vents and how they are installed.

LOOP AND CIRCUIT VENTS

Loop and circuit vents are used on a line of fixtures. A *loop vent* is used in single-story houses or on the top floor of a multistory building. In the loop vent, the vent branch

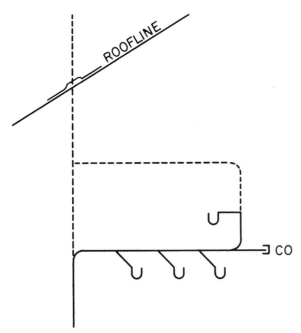

Fig. 26-1 Loop vent

goes back into the stack vent at a point above all drain inlets, figure 26-1.

The horizontal branch vent pipe on the loop vent must pitch back to the traps. This prevents condensation from collecting at any point. Condensation also collects in drops along the top of horizontal vent pipes. The branch drain must continue full size to the last fixture connection.

A *circuit vent* is shown in figures 26-2 and 26-3. In each case, it is limited in length so that the total pitch of the drain does not exceed 18 inches. The number of fixtures is limited by the size of the pipe.

The vertical vent at the end of the drain must be washed out by having a fixture connected at that point. A cleanout plug must be placed at the end of the drain.

Relief vents are sometimes installed to relieve air pressure on branch waste near the stack, see figure 26-3. In tall buildings relief vents, also called *yoke vents,* are installed at 10-story intervals from the top to relieve air pressure, figure 26-4, page 112.

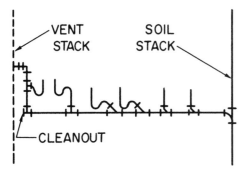

Fig. 26-2 Circuit vent

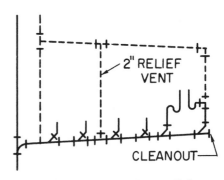

Fig. 26-3 Circuit vent. Note relief vent.

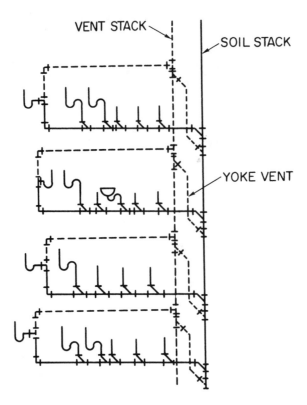

Fig. 26-4 Yoke vent

REVIEW QUESTIONS

Multiple Choice

Select the best answer for each question.

1. A vent line which vents a number of fixtures and then goes back into the stack is known as a

   a. crown vent.   c. stack vent.
   b. circuit vent.   d. loop vent.

2. A vent which goes between the soil stack and the vent stack at ten branch intervals is known as a

   a. stack vent.   c. circuit vent.
   b. yoke vent.   d. bleeder vent.

3. Why does a branch vent pitch back toward the traps?

   a. So the discharge from a fixture goes in the right direction
   b. So scale falls down the pipe
   c. So condensation does not collect
   d. None of the above

4. What type of vent connects to the fixture farthest from the stack in figure 26-1?

   a. A loop vent
   b. A circuit vent

   c. A yoke vent
   d. A continuous vent

5. Why is the last fixture in figure 26-2 connected to the vertical vent pipe?

   a. To form a back vent
   b. So the heel of the loop is washed out
   c. To provide a relief vent
   d. To relieve back pressure

6. What does the CO in figure 26-1 stand for?

   a. Cleanout
   b. Closet outlet

   c. The artist's initials
   d. None of the above

7. Loop vents are commonly used in

   a. apartment projects.
   b. continuous venting.

   c. single-story buildings.
   d. split-level homes.

8. How many stories from the top should the first yoke vent be placed?

   a. 3, including the top
   b. 9, including the top

   c. 10, including the top
   d. 15, including the top

9. Why is it incorrect to connect a yoke vent higher on the soil stack than it is on the vent stack?

   a. The yoke vent will not function.
   b. Water will go down the vent stack.
   c. Scale will not fall out of the vent.
   d. None of the above.

UNIT 27 SOVENT DRAINAGE SYSTEMS

OBJECTIVES

After studying this unit, the student will be able to:

- explain how a SOVENT drainage system works.

- discuss why a SOVENT drainage system costs less than a two-pipe system.

DRAINAGE SYSTEMS

In any drainage system, traps prevent sewer gas from entering a building. Because traps must be simple, absolutely reliable, and have no moving parts, the water-filled P or S trap is used.

The seal in a water-filled trap is quite fragile. A small pressure from above (siphonage) or from below (back pressure) will upset the trap seal. The design of all drainage systems must therefore insure that pressure within the system is balanced with atmospheric pressure.

In conventional two-pipe systems, balance is maintained by separate systems for the movement of air and waste. The SOVENT drainage system, however, accomplishes this by using only one stack in a self-venting system.

SOVENT DRAINAGE SYSTEM

The SOVENT drainage system is designed to simplify drainage, waste, and vent piping in multistory buildings. It was developed in Switzerland as a less expensive alternative to the two-pipe system commonly used in the United States.

The SOVENT system uses one pipe and specialized fittings to prevent the stack from becoming completely filled at any point,

figure 27-1. If the cross section of the stack is never completely filled, then air is free to move within the same stack as the waste. This will permit the air to equalize pressures within the stack.

The SOVENT design:

- slows down the speed of the flow at each floor.

- causes a turbulence at each floor which introduces air into the waste.

- removes the air as the waste reaches the base of the stack.

- prevents positive pressures from developing.

The result is a single stack that is self-venting with the fittings balancing pressures throughout the system. SOil stack and VENT combine into a single SOVENT stack.

The SOVENT design utilizes two unique fittings: the *aerator* and the *deaerator*. These special fittings are the basis for the self-venting features of the SOVENT.

THE AERATOR

The aerator fitting has a built-in off-set which serves to slow down the flow,

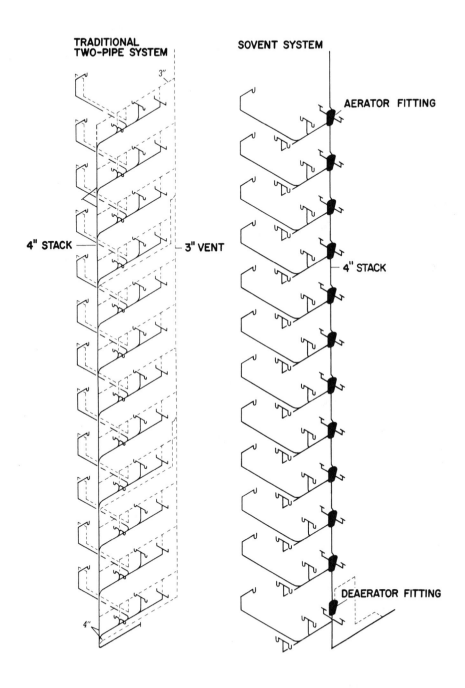

Fig. 27-1 Compare the two-pipe and SOVENT drainage systems.
Note the SOVENT system requires less material — a cost-saving feature.

figure 27-2. It has a mixing chamber, one or more branch inlets, one or more waste inlets for connection of smaller waste branches, a baffle in the center of the chamber, and the stack outlet at the bottom of the fitting.

The aerator provides a chamber where the flow of soil and waste from the horizontal branches can unite smoothly with the air and liquid already flowing in the stack. The aerator fitting does this so efficiently that the stack cannot become completely filled and cause pressure changes.

Aerator fittings are placed in the stack at every floor where there is a substantial branch. Where there is no substantial branch line, a double in-line offset is installed in its place, see figure 27-2. The purpose of the offset is to slow down the flow, just as the aerator does.

THE DEAERATOR

The deaerator is placed near the base of the stack. It prevents waste from slowing down at the base and forming a slug of water which would cause pressure changes. For the same reason, a deaerator is placed at every offset in the stack, except where there is a double in-line offset, figure 27-3.

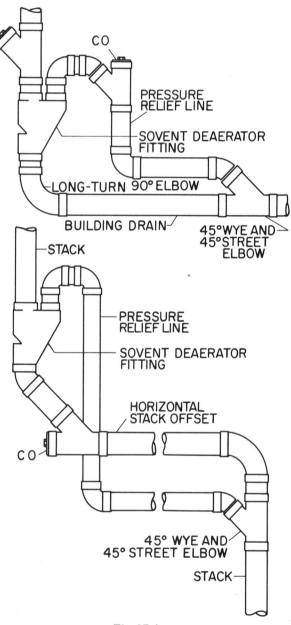

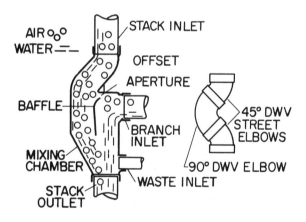

Fig. 27-2 SOVENT aerator

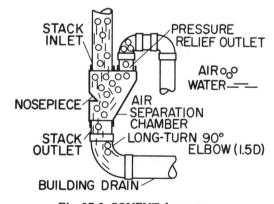

Fig. 27-3 SOVENT deaerator

Fig. 27-4

The deaerator functions along with the aerator above it to make a single stack self-venting. It consists of an air separation chamber with an internal projection, a stack inlet, a pressure relief outlet at the top, and a stack outlet at the bottom, figure 27-4. In practice, some of the air in the descending slug is ejected into the relief line. This is led back into the house drain at a point at least 4 feet from the base of the stack, figure 27-5.

BRANCH LINES

The SOVENT design does not require back venting until the horizontal run exceeds 27 feet. The long horizontal distances without

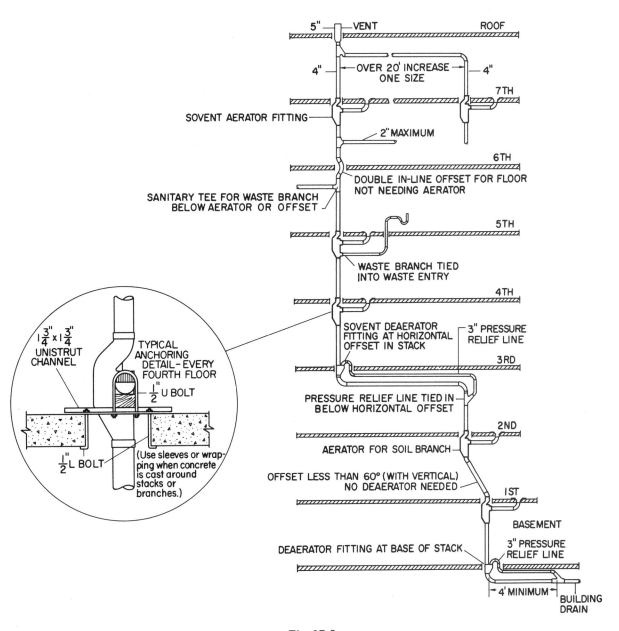

Fig. 27-5

a separate vent permitted by the SOVENT design are achieved by oversizing the branches. This prevents them from ever being completely filled.

PLUMBING CODE REGULATIONS

In many places, the plumber must get special permission from the sewer authority to use the SOVENT system. This system has, however, been installed in more than one hundred apartment and office buildings in the United States alone.

Additional information about SOVENT plumbing may be obtained from:

Copper Development Association, Inc.
405 Lexington Avenue
New York, New York 10017

REVIEW QUESTIONS

Multiple Choice

Select the best answer for each question.

1. What basic plumbing fitting is used to keep sewer gas from entering a building?

   a. An aerator
   b. A long-turn elbow
   c. A trap
   d. A backwater valve

2. A SOVENT drainage system costs less than a traditional plumbing system because

   a. it does not use traps.
   b. it uses less-expensive materials.
   c. it is only used on one-story houses.
   d. it requires less piping.

3. SOVENT plumbing is a

   a. single-stack system.
   b. double-stack system.
   c. triple-stack system.
   d. none of the above.

4. The aerator prevents the stack from being completely

   a. submerged.
   b. air-bound.
   c. filled.
   d. drained.

5. The aerator is found at

   a. every floor.
   b. every floor that has a substantial branch.
   c. the base of the stack.
   d. the main drain.

6. The purpose of the double in-line offset is to

   a. slow down the flow.
   b. add air to the waste.
   c. separate the air from the water.
   d. vent the line.

7. The deaerator is found

   a. at the top of the stack.
   b. at the base of the stack only.
   c. at every offset.
   d. at the base and at every offset.

8. The relief pipe from the deaerator goes into the top of the main house drain _____ from the base of the stack.

   a. 2 feet   c. 6 feet
   b. 4 feet   d. 8 feet

9. The pipe which is connected to the relief outlet of the deaerator is then connected to the

   a. aerator.   c. house drain.
   b. vent stack.   d. roof vent.

10. By oversizing branch lines in the SOVENT system,

    a. the pipe will be stronger.
    b. the pipe will not fill completely.
    c. the trap will be eliminated.
    d. the trap will empty completely.

UNIT 28 CROSS CONNECTIONS

OBJECTIVES

After studying this unit, the student will be able to:

- describe how and where cross connections are used.

- discuss the causes and dangers of back siphonage through water pipes.

CROSS CONNECTIONS

Cross connections are direct connections between a water supply and a system carrying unsafe water. Preventative measures may now be taken to prevent this dangerous situation.

Back siphonage through water pipes from drains is caused by failure or a drastic reduction of the water pressure. Pressure reduction may be caused by the sudden flushing of street mains, fire engines pumping, breaks in mains, or improper closing of street valves. Inside buildings, it is caused by meters,

strainers, pressure regulators, or undersized piping. Adding fixtures to existing pipelines results in an excessive pressure drop.

When any one of these conditions occurs while a fixture is clogged, the water falling back in a water riser creates a partial vacuum which draws sewage into the water system, figure 28-1.

In tall buildings, siphonage may be severe enough to draw water from a lavatory into a faucet even though the faucet nozzle is above the water surface. This may be corrected by elevating the nozzle to at least 1 1/2 inches above the rim of the fixture, figure 28-2.

A reliable vacuum breaker installed between flushometer valves and bowls usually prevents this condition in toilets, figure 28-3.

In some large buildings, untreated river water may be used to flush toilets, test tanks, and other fixtures. A separate system of pipe

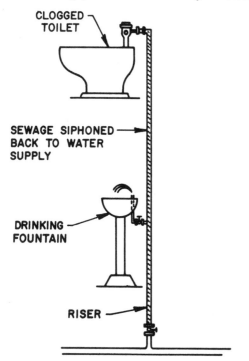

Fig. 28-1

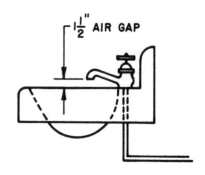

Fig. 28-2

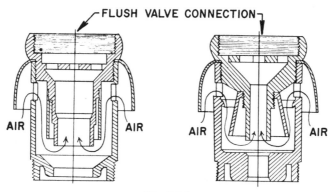

Fig. 28-3

is required, but the two systems are cross-connected in an emergency such as a fire. The danger occurs when someone unknowingly opens the valve between the pure and the unsafe water supplies. If this occurs, many people may become dangerously ill. *Note:* Check with your local code. Using untreated river water is strictly forbidden in some areas of the United States.

All pipelines containing unsafe water are painted red. The valve must be labeled and check valves installed. This prevents unsafe water from entering the safe water supply.

Overhead drain lines should be inspected periodically to prevent sewage water from leaking into tanks containing food. Since fixtures below sewer level are subject to stoppage and flooding, backwater valves should be installed. When water supply systems are designed, the mains must be large enough to maintain 15 psi of pressure at the highest fixture in addition to the probable flow.

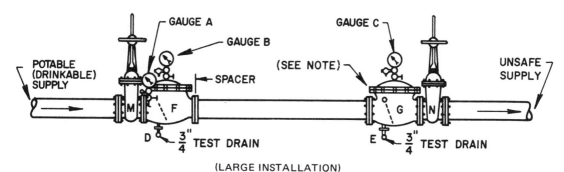

Fig. 28-4

REVIEW QUESTIONS

Multiple Choice

Select the best answer for each question.

1. A cross connection in a water supply system could be

    a. when two waterlines cross.
    b. a connection between the hot and the cold mains.
    c. a connection between a fire main and a drinking water supply.
    d. none of the above.

2. The danger of a cross connection is that

    a. people may be seriously poisoned.
    b. a loss of water pressure may occur.
    c. water will be wasted.
    d. sewer gas may enter the house.

3. What is the most common cause of back siphonage?

    a. A loss of trap seals
    b. A lack of knowledge on the part of the tenant
    c. Lavatories with no air gap
    d. A reduction in water pressure

4. Why could corroded pipes cause back siphonage?

    a. Corroded pipes are the same as undersized piping.
    b. Scale could plug the fixture.
    c. Corroded pipes are more likely to be cross-connected.
    d. They may sag.

5. If water from the clogged toilet in figure 28-1 is siphoned back into the water main, where will it be drawn when the pressure returns to normal?

    a. Into the water main   c. Back into the toilet
    b. Into the water meter   d. Distributed to the fixtures

6. What force pushes the water from the toilet into the water main?

    a. Atmospheric pressure   c. Inertia
    b. Electromagnetic forces   d. Osmosis

7. How are pipes containing unsafe water distinguished from those containing safe water?

    a. Examine the blueprints carefully.
    b. Unsafe water pipes are colored blue.
    c. Unsafe water pipes are colored red.
    d. None of the above.

8. What pressure should be designed at the fixtures of a water system?

    a. 8 pounds   c. 40 pounds
    b. 15 pounds   d. 55 pounds

9. What precaution should be taken when drainpipes are located over food supplies?

    a. They should be painted red.
    b. They should be periodically inspected.
    c. They should be covered with an insulating material.
    d. Sheet metal should be placed above them.

10. What is an air gap?

    a. A period of time when there is a pressure loss
    b. A device which is installed on a flushometer
    c. A vertical space between the water outlet and the flood rim of a fixture
    d. A special fitting for all fixtures

UNIT 29 FIXTURE AND GREASE TRAPS

OBJECTIVES

After studying this unit, the student will be able to:

- describe the types of fixture traps and where they are used.
- explain why and where grease traps are used.

FIXTURE TRAPS

Sewers, drains, and waste pipes contain sewer gases such as hydrogen sulphide, methane, and carbon dioxide. They also contain germs and disease-carrying vermin. Traps are installed on each main drain and at each fixture to prevent the gases and vermin from entering buildings.

A *trap* is a receptacle placed in the waste pipe from a plumbing fixture. It holds a quantity of water which blocks the passage of vermin and sewer gases. It does not restrict the flow of waste. The trap must be installed as close to the fixture as possible. It should never be more than 24 inches below the outlet of the fixture.

Most trap patterns are the S type, figure 29-1. The S type is the same diameter throughout and has a smooth self-cleaning action. The *1/2 S trap*, or *P trap* as it is commonly called, is the most frequently used fixture trap. The *full S trap* is usually prohibited by plumbing codes because it presents obstacles to the proper venting of the trap. The *drum trap* is also generally prohibited because it does not have a good self-cleaning action.

The *seal* of a trap is the water between the dip and the outlet, figure 29-2. If this water is lost, the effectiveness of the trap is destroyed. A safe seal is from 2 1/2 to 4 inches deep.

Traps on individual fixtures should be the same size as the waste pipe. All traps should have smooth interiors, no internal partitions, depend on no movable parts to form a seal, and have a safe depth of seal.

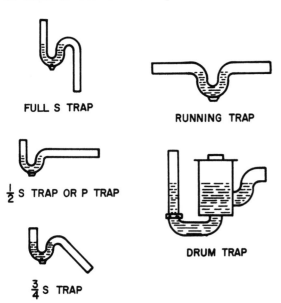

Fig. 29-1 Types of traps

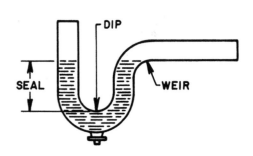

Fig. 29-2

Fixture	Trap Size	Fixture	Trap Size
Bathtub	1 1/2″	Pantry or soda fountain	1 1/2″
Drinking fountain	1 1/4″	Restaurant or hotel sink	2″
Floor drain	3″ or 4″	Shower	2″
Kitchen sink	1 1/2″	Sitz bath	1 1/2″
Laundry tray	1 1/2″	Urinal	2″
Lavatory	1 1/4″	Water closet	2 1/4″

Fig. 29-3

Figure 29-3 lists the sizes of traps for certain fixtures.

Antisiphon traps are obsolete. Today, engineers design plumbing systems so they prevent siphonage and back pressure. They do not depend upon a special trap to do an impossible task. To maintain the seal of a trap, the pressure on each side of the seal must be within one ounce of being equal. This is accomplished by proper venting and correct pipe sizes.

A 2 1/2-inch seal will only withstand about 1.5 ounces per square inch of pressure. Deeper seals cannot be used as the trap would not be self-cleaning.

GREASE TRAPS

A *grease trap* is a receptacle placed in the waste pipes of sinks to separate and retain grease from the water. These traps are used mostly in hotels and restaurants because of the quantity of grease in the waste water. While the water is hot, the grease is in liquid form and floats on top of the water. The grease thickens when cooled and sticks to the sides of water pipes, clogging them.

In rural sewage systems, the grease clogs the joints of loose cesspool and subsoil irrigation tile. Grease becomes as hard as soap in time and will eventually fill the pipe.

Figure 29-4 shows an example of a grease trap with baffle plates. The trap should be large enough to hold twice the capacity of the fixture. This allows the contents to cool before the next flush. The baffle plates are placed so they separate the grease. The grease, since it is lighter than water, rises to the top. The outlet is taken from the bottom to prevent grease from passing into the waste pipe. It has a large cleanout cover held in place by thumbscrews which may be removed for cleaning.

Grease traps which use a potable (drinkable) water supply for cooling are illegal.

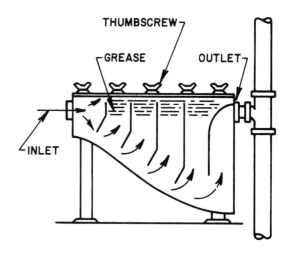

Fig. 29-4 Grease trap

REVIEW QUESTIONS

Multiple Choice

Select the best answer for each question.

1. The purpose of a fixture trap is to

   a. keep sewer gases and vermin from entering a building.
   b. save water.
   c. provide a firm attachment from the fixture to the wall.
   d. vent the fixture.

2. A trap may not be installed more than _____ below a fixture.

   a. 16 inches
   b. 20 inches
   c. 24 inches
   d. 48 inches.

3. An S trap has a good

   a. sewer resistance.
   b. hydraulic gradient action.
   c. corrosion resistance.
   d. self-cleaning action.

4. How are sewer gases prevented from entering buildings through waste pipes?

   a. By proper venting
   b. By the installation of graded piping
   c. By trap seals
   d. By the use of vacuum breakers

5. What happens if the water in a trap evaporates below the dip?

   a. Corrosion occurs along the top of the trap arm.
   b. Offensive odors enter the building.
   c. The fixture is not able to hold water.
   d. The cleanout plug will leak.

6. What is the minimum safe depth of seal for a trap?

   a. 2 1/2 inches
   b. 5 inches
   c. 5 inches.
   d. 24 inches

7. One problem with a drum trap is that

   a. it cannot be built strong enough.
   b. it must be cleaned out from the top.
   c. it does not have good self-cleaning action.
   d. none of the above

8. Where are grease traps found?

   a. In a laundry tub drain
   b. In a garage drain
   c. At the base of a soil stack
   d. In the waste pipe from a kitchen sink

9. How large should the grease trap be?

   a. Large enough to hold all of the fixture's discharge
   b. 20 gallons
   c. 12 gallons
   d. Twice the capacity of the fixture

Plumbing Sketch

Make a sketch of a P trap and indicate the seal.

UNIT 30 GARAGE SAND TRAPS

OBJECTIVE

After studying this unit, the student will be able to:

- explain how garage sand traps work.

GARAGE SAND TRAPS

The drainage from garages is handled differently than the drainage from homes and other buildings. This is because the drainage from garages contains sand, gasoline, and oil. The sand which falls from or is washed from

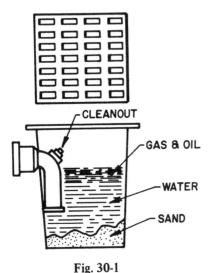

Fig. 30-1

cars, for instance, may clog a regular drain or sewer.

The purpose of the garage sand trap is to prevent sand, gasoline, and oil from entering regular sewer drains. It consists of a catch basin 18 inches deep with a cover. The outlet is taken from an elbow turned down below the water surface, figure 30-1. The top of the trap is level with the cement floor. Since gas and oil are lighter than water, they float to the surface and do not enter the drain. The sand settles to the bottom. These traps should be cleaned periodically.

In large public garages where several floor drains are installed, a brick or concrete sand trap is used instead of the small garage sand trap, figure 30-2. This type of trap is constructed with two compartments. The overflow from each is taken off with a sweep bend which is turned down. This prevents the gas and oil, which float on the top of the

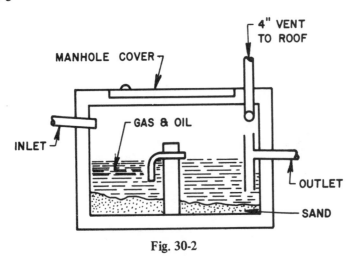

Fig. 30-2

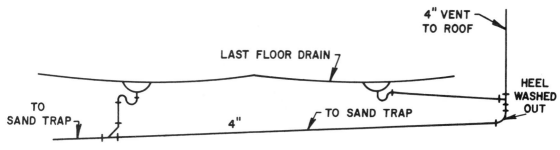

Fig. 30-3

water, from being carried into the sewers. The dirt and sand are retained in the first chamber. Should the gas and oil overflow into the second chamber, they are checked there by the inverted outlet.

The sand trap, which is vented separately to the roof, has a loose steel cover for cleaning.

The outlet is connected to the main drain at any convenient location. The end of this separate drain must be vented to the roof, and the last floor drain must run to this stack to wash out the rust scale, figure 30-3. This sand trap can only be effective if it is periodically cleaned.

REVIEW QUESTIONS

Multiple Choice

Select the best answer for each question.

1. What happens if oil and water are poured into a bucket?

   a. The oil sinks to the bottom of the bucket.
   b. The oil and water slowly mix together.
   c. The oil floats on top of the water.
   d. None of the above.

2. What happens if the sand is not cleaned out of the sand trap periodically?

   a. The vent becomes clogged.
   b. The old sand becomes clogged with oil.
   c. The pipes become worn from the passage of sand.
   d. The sand builds up until it closes the outlet of the trap.

3. The purpose of the garage sand trap is to

   a. prevent sand, gas, and oil from entering the sewer.
   b. use the sand to filter out gas and oil.
   c. reclaim sand for later use.
   d. prevent flooding the garage floor.

4. Why must garage sand traps have heavy covers?

   a. To support the walls of the trap
   b. So that vehicles will not break it
   c. So that unauthorized persons will not tamper with it
   d. Because sand is heavy

5. In figure 30-4, which location indicates, the inlet of the sand trap?

    a. 1   c. 3
    b. 2   d. None of the above

6. In figure 30-4, which location indicates the vent to the sand trap?

    a. 1   c. 3
    b. 2   d. None of the above

7. In figure 30-4, which location indicates the outlet to the sand trap?

    a. 1   c. 3
    b. 2   d. None of the above

8. What is the purpose of the cleanout in figure 30-1?

    a. To clean out the sand in the trap
    b. To clear a stoppage from the drain line
    c. To remove the gas and oil without removing the cover
    d. To draw in fresh air

9. Why must the drain from the last floor drain be run to the stack rather than the drain?

    a. To wash out the heel of the stack
    b. To avoid crowding the branch fittings
    c. So that the main drain may be run higher
    d. None of the above.

10. Why is it necessary to vent a large sand trap?

    a. To prevent air-locking   c. To vent gasoline fumes
    b. To promote aerobic action   d. To provide a surge outlet

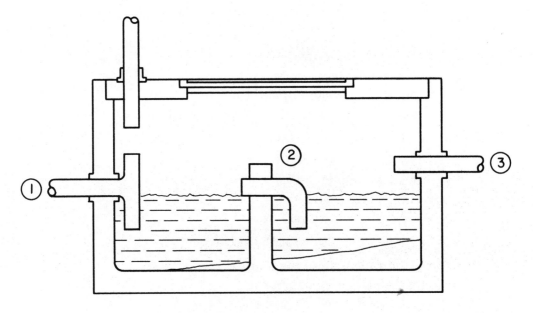

Fig. 30-4 Drawing for Unit Review

UNIT 31 RAIN LEADERS

OBJECTIVES

After studying this unit, the student will be able to:

- discuss the purpose of rain leaders.
- explain where and how rain leaders are installed.

RAIN LEADERS

A *rain leader* is a pipe which carries rainwater from the roof or gutter of a structure to the correct disposal point.

Running rainwater into sewage systems is not allowed in today's ecologically-minded society. Although there are still many combination sewers in use, these are being replaced by separate storm water systems. The argument that a heavy storm may flush out the sewage system is no longer valid. This is because most systems today run well above their normal capacity. The discharge of surface water into the system does little other than overload the system. However, material which is difficult to remove may enter the system in this manner. Most codes require a separate system of piping for rain leaders and surface drains.

Outdoor, above ground rain leaders are usually installed by the roofing contractor. If the underground portion discharges at the curb, there is no need for a trap.

Indoor rain leaders are subject to the same regulations as drainpipes. They must have a cast-iron, deep-seal trap with a cleanout which is accessible for cleaning. Indoor rain leaders have the advantage of not freezing since they are located in heated buildings. In cold buildings, such as warehouses, they are protected from freezing by steam pipes or thermostatically-controlled, electric heat tapes.

The connection to the roof is sometimes made of lead and then wiped and caulked into the soil pipe or galvanized pipe. Figure 31-1 shows one method. A piece of lead pipe is wiped to the sheet lead and a ferrule and caulked into the soil pipe.

Figure 31-2 shows another method in which cast-iron pipe may be used. The horizontal run of pipe tied into the vertical leader provides for some expansion in any direction. Cast-iron strainers are used to keep leaves and other debris from closing the drain. The strainer should extend at least 4 inches above the surface of the roof.

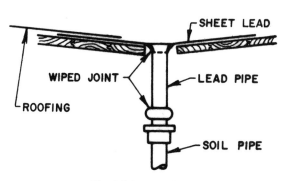

Fig. 31-1 Wiped joint

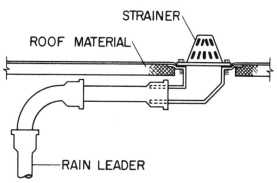

Fig. 31-2 Expansion joint

REVIEW QUESTIONS

Multiple Choice

Select the best answer for each question.

1. Rainwater should not be discharged into a sewage treatment plant because

   a. it overloads most plants.
   b. it wastes valuable water resources.
   c. it creates excessive pressures within the city sewers.
   d. it causes corrosion.

2. Where must rainwater be discharged?

   a. Into the correct point of disposal
   b. Into a private septic system
   c. Into the correct storage system
   d. Below ground level

3. What is a sewer for rainwater called?

   a. Field tile
   b. A sanitary sewer
   c. A house sewer
   d. A storm sewer

4. The purpose of a rain leader is to

   a. keep the roof from leaking.
   b. carry rain from the roof.
   c. protect passersby from rain runoff.
   d. vent the sanitary sewer.

5. How do indoor rain leaders differ from outdoor rain leaders?

   a. Lighter materials may be used indoors.
   b. Indoor rain leaders are never heated.
   c. Indoor rain leaders require traps.
   d. None of the above.

6. Why is it necessary to use a deep seal trap on a rain leader?

   a. Because the trap may dry out during dry spells
   b. To protect the trap from freezing
   c. To protect the city sewer from overloading
   d. To collect roof runoff

7. How are leaders in unheated buildings protected?

   a. By adding antifreeze to the traps
   b. By steam lines or electrical heating tapes
   c. By running them within the walls
   d. By oversizing them

8. Lead roof drains are put together from sheet lead and lead pipe by means of

   a. the wiped joint.   c. roofing cement.
   b. lead welding.   d. brazing.

9. How far above the roof should the strainer extend?

   a. Even with the roof
   b. 1 1/2 times the leader diameter
   c. 1 inch above the high water mark
   d. At least 4 inches

10. The purpose of a strainer on a roof drain is to

   a. keep the rainwater pure.
   b. keep debris from clogging the leader.
   c. slow down the flow during severe storms.
   d. trap vermin within the system.

UNIT 32 SUMP PUMPS AND CELLAR DRAINERS

OBJECTIVES

After studying this unit, the student will be able to:

- explain how water may be removed from a cellar.

- describe the installation and operation of a sump pump.

SUMP PUMPS AND CELLAR DRAINS

In localities where the ground water is close to the surface, basements are likely to be wet, especially in rainy weather. This can be a real health problem.

The depression in a basement floor where water collects is known as a *sump*. In new construction, this condition can be remedied by laying a system of field tile or perforated drainpipe around the outside of the basement wall. The tile is then connected to a *sump pump* in the basement and below the floor, figure 32-1. Elevator pits are drained in this manner. The field tile is covered with broken stone to insure better drainage.

To remedy such a condition in an existing building:

1. Build a sump under the basement floor and place holes in the sides of it.

2. Extend field tile around the outside of the cellar wall and connect it to the sump.

3. Install an automatic cellar drainer to pump the water up to the drain if it will not run out by gravity.

There are several types of cellar drainers. Older, now obsolete types used water, air, or steam pressure through a jet to siphon the water out. This type of siphon raised water 1 foot for every 4 psi of pressure. It wasted

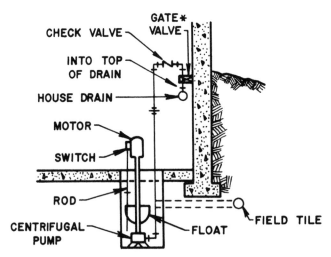

*When connected to a house drain, some codes require both the check valve and gate valve.

Fig. 32-1

considerable water. It also posed a constant threat of forcing contaminated water back into the potable (drinkable) water lines when connected to it.

One of the newer cellar drainers consists of an electrically-driven, vertical centrifugal pump. It is controlled by a float and is more efficient than older styles. The motor is installed above the highest possible water level so that it will remain dry. As the water rises in the sump, the float pushes the rod up and turns on the switch, starting the motor. As the water recedes, the float pulls the rod down and shuts off the motor.

Another, still newer type of cellar drainer is a completely submersible unit which operates underwater. The unit, therefore, is never flooded out and operates until the current source is cut off or the desired water level is reached. It can be lowered into a flooded cellar or pit to pump out water through a flexible pipe or hose. The switch, which is a sealed unit, is more effective than the rod-and-float type.

Care must be taken to avoid electrical shock when working around cellar drainers since they are always located in damp areas. Be certain that the unit is properly grounded.

Unions should be placed on all pipes which lead to the pump. This makes removal of the pump for repairs when the sump is full of water easier. The discharge from the cellar drainer should extend to a safe terminal, such as a storm drain or suitable dry well. This discharge should not connect to the sanitary sewer. It should also be protected by a suitable check valve to prevent the backflow of surface water into the sump.

REVIEW QUESTIONS

Multiple Choice

Select the best answer for each question.

1. Ground water enters basements when
   a. someone leaves the plug out.
   b. the water table rises during rainy spells.
   c. fixtures located in the basement overflow.
   d. cellar windows are left open.

2. The name for the depression in a cellar floor where water collects is called a
   a. laundry tub.
   b. cesspool.
   c. sump.
   d. well casing.

3. Field tile is used to
   a. carry excess water out to the field.
   b. carry water from around the foundation into the sump.
   c. make basement floors waterproof.
   d. provide grounding.

4. Perforated pipe used to drain water from foundation walls is surrounded with crushed stone
   a. because crushed stone attracts water.
   b. to give the pipe the proper fall.
   c. to keep the pipe from sagging.
   d. to keep the holes from plugging with dirt.

5. Two types of cellar drainers are

    a. the field tile type and the perforated drainpipe.
    b. the sump and the siphon jet.
    c. the floor drain and the siphon.
    d. the vertical centrifugal and the submersible.

6. Where should the motor in the vertical centrifugal cellar drainer be placed?

    a. In the sump
    b. Above the highest possible water level
    c. Within the pump unit itself
    d. Near the fuse box

7. On the vertical centrifugal-type pump, the motor is _____ when the float rises to its highest setting.

    a. turned on   c. pressure primed
    b. turned off   d. shorted out

8. How high will 32 psi of water pressure raise water through a siphon-type cellar drainer?

    a. 4 feet   c. 12 feet
    b. 8 feet   d. 16 feet

9. What conditions may exist in the drainpipe which makes a check valve necessary in the discharge pipe from the sump pump?

    a. No vent   c. A backed-up drain
    b. A loss of trap seal   d. A severe vacuum

10. Unions are placed on all pipes leading to the pump so that

    a. electrical shock can be avoided.
    b. surface water does not backflow into the sump.
    c. the sump can be cleaned out.
    d. the pump can be removed when the sump is full of water.

UNIT 33 BLOW-OFF TANKS

OBJECTIVE

After studying this unit, the student will be able to:

• explain why blow-off tanks are needed in drainage systems.

BLOW-OFF TANKS

To remove mud, loose scale, and mineral deposits, large steam boilers are *blown down*. A valve at the bottom of the boiler is opened, while pressure is maintained, and deposits are removed. Since the water is at a high temperature (about 215 to 250 degrees) which would damage the drainpipes, it must be discharged into a blow-off tank, figure 33-1.

The purpose of the blow-off tank is to cool the water before it enters the drainage system. When the valve in the blow-off pipe is opened, the water supply valve is also opened. Cold water is then sprayed on the hot water, thereby lowering its temperature.

The steam is relieved through the vent. Cooled water from the bottom of the tank is discharged into the drainpipe. The vent is carried to the roof, which prevents the tank from being siphoned.

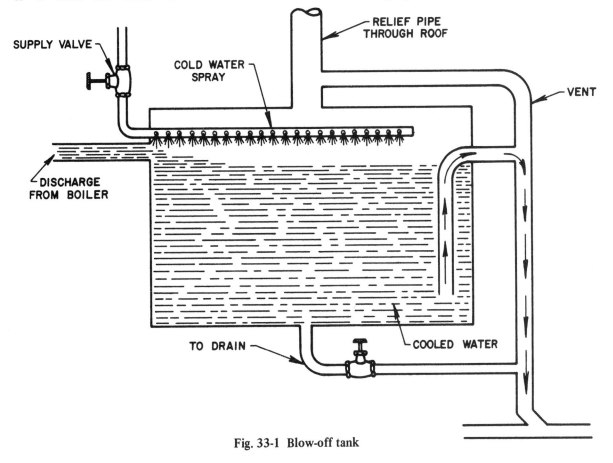

Fig. 33-1 Blow-off tank

REVIEW QUESTIONS

Multiple Choice

Select the best answer for each question.

1. What is the approximate temperature of the water in a steam boiler?

   a. 72 to 112 degrees
   b. 135 to 160 degrees
   c. 195 to 212 degrees
   d. 215 to 250 degrees

2. What will happen to ABS drainpipe if high-temperature water is discharged into the pipe?

   a. Nothing, ABS is temperature certified.
   b. The pipe will soften.
   c. The joints will come apart.
   d. Cracks will appear.

3. The purpose of a blow-off tank is to

   a. remove minerals from boiler water.
   b. reduce the temperature of blow-off water.
   c. prevent grease and oil from entering the sanitary sewer.
   d. provide boiler feed water.

4. Into which part of the tank is hot water discharged?

   a. The relief pipe
   b. The bottom
   c. The vent bypass
   d. The top

5. From which part of the tank does the cold water run to the drain?

   a. From the top outlet
   b. From a point 1/4 of the way up from the bottom
   c. From the vent overflow
   d. From the bottom

6. The hot water is cooled by

   a. holding it in the tank for at least 1/2 hour.
   b. air circulating through the vent.
   c. a spray of cold water.
   d. refrigeration.

7. The purpose of the relief pipe is to

   a. keep the vent from the blow-off tank separate from all others.
   b. provide an escape for the steam.
   c. provide a safe terminal for the safety valve.
   d. keep temperatures high.

8. On the boiler, the blow-down valve is attached to

   a. the boiler feed valve.
   b. the steam main.
   c. the relief valve.
   d. the bottom.

9. When the discharge valve from the boiler is opened, the _____ should also be opened.

   a. drain valve   c. supply valve
   b. vent pipe   d. relief valve

10. While the boiler is being blown down, _____ is maintained.

    a. pressure   c. security
    b. humidity   d. temperature

SECTION 5

Hot Water Supply

UNIT 34 BRITISH THERMAL UNIT AND THE EXPANSION OF WATER

OBJECTIVES

After studying this unit, the student will be able to:

- describe the British thermal unit and how it relates to plumbing.
- discuss the expansion of water and the pressures it produces.

THE BRITISH THERMAL UNIT

Heat may be measured in degrees or in British thermal units. A *British thermal unit* (btu) is the quantity of heat necessary to raise the temperature of 1 pound of water 1 degree. A pint of water is approximately 1 pound. Therefore, 1 Btu will increase a pint of water from 54 to 55 degrees Fahrenheit. Likewise, 10 Btu will raise 1 pound of water 10 degrees and 25 Btu will raise 25 pounds of water 1 degree.

To estimate the quantity of heat in Btu required to raise water a certain number of degrees, multiply the number of pounds of water by the number of degrees it is to be raised.

Example. Find the number of Btu necessary to heat 100 pounds of water 40 degrees.

Solution. 100 lb × 40° = 4000 Btu

If the quantity of water is given in gallons, multiply by 8.33 pounds (weight of a gallon). If the quantity is given in cubic feet, multiply by 62.5 pounds (weight of a cubic foot). If the two temperatures are given, find the temperature difference by subtracting.

Example. Find the number of Btu necessary to raise 30 gallons of water from 40 degrees to 160 degrees.

Solution. 8.33 lb × 30 gal = 249.9 lb
160° – 40° = 120° temperature difference
249.9 lb × 120° = 29,988 Btu

EXPANSION OF WATER

The expansion of water differs from that of metals. Metals expand as long as heat is applied and contract as long as heat is extracted. Water, however, has a point of maximum density. At 39.2 degrees Fahrenheit, water reaches a point where it can no longer contract. Water expands when heated above this temperature. It also expands when

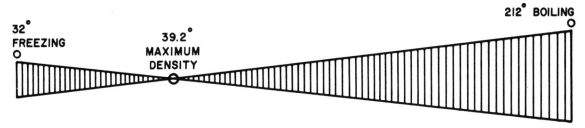

Fig. 34-1

cooled below 39.2 degrees Fahrenheit. Water expands about 1/26 of its volume when heated to the boiling point, figure 34-1.

One cubic inch of water expanded to steam will fill a cubic foot of space, expanding 1728 times its original volume. Water converted to steam will drive a locomotive. Ice exerts a pressure of 33,000 pounds per square inch. Water converted to ice will most likely break a pipe.

The weight of a cubic foot of water becomes lighter as it is heated or cooled below 39.2 degrees Fahrenheit, figure 34-2.

Note: Realizing the tremendous pressure exerted by expanding water, a reliable safety valve must be placed on all water-heating apparatus.

Ice	at	32°	weighs	62.418	lb.	per	cubic	foot
Water	at	39.2°	weighs	62.425	lb.	per	cubic	foot
Water	at	60°	weighs	62.372	lb.	per	cubic	foot
Water	at	160°	weighs	60.991	lb.	per	cubic	foot
Water	at	212°	weighs	59.760	lb.	per	cubic	foot

Fig. 34-2

REVIEW QUESTIONS

Multiple Choice

Select the best answer for each question.

1. The British thermal unit is a measurement of

   a. the temperature of an object.

   b. the quantity of heat necessary to change the temperature of a substance.

   c. the number of degrees necessary to change the state of a substance.

   d. the seventh law of thermodynamics.

2. How is metal affected by an increase or decrease in temperature?

   a. Metal expands on increase and contracts on decrease.

   b. Metal expands on decrease and contracts on increase.

   c. Metal expands on decrease and remains the same on increase.

   d. Temperature does not affect metal.

3. How many Btu are necessary to raise 20 pounds of water 2 degrees?
   a. 2 Btu
   b. 10 Btu
   c. 20 Btu
   d. 40 Btu

4. How many Btu will raise 18 pounds of water from 23 degrees to 70 degrees?
   a. 414 Btu
   b. 846 Btu
   c. 1260 Btu
   d. 1674 Btu

5. How many Btu will raise 25 gallons of water 35 degrees?
   a. 10 Btu
   b. 875 Btu
   c. 7288.75 Btu
   d. 54687.5 Btu

6. How many Btu will raise 40 gallons of water from 40 degrees to the boiling point?
   a. 1600 Btu
   b. 8480 Btu
   c. 19992 Btu
   d. 57310.4 Btu

7. How many Btu will heat 11 cubic feet of water from 42 degrees to 140 degrees?
   a. 1078 Btu
   b. 1540 Btu
   c. 8979.74 Btu
   d. 67375 Btu

8. How many Btu will heat the water in a rectangular tank 8 feet long, 4 feet wide, and 5 feet deep from 40 degrees to 135 degrees?
   a. 640 Btu
   b. 21600 Btu
   c. 950000 Btu
   d. 1350000 Btu

9. Water weighs the most at
   a. 32 degrees.
   b. 39.2 degrees.
   c. 160 degrees.
   d. 212 degrees.

10. If a can full of water is placed on a lighted gas burner, what will happen?
    a. The water will expand and overflow the can.
    b. There will be no increase or decrease in volume.
    c. The contraction of the contents will collapse the sides of the can.
    d. none of the above

11. What will happen if a pipe is filled with water, plugged tightly, and then frozen solid?
    a. The pipe will stretch in length.
    b. The pipe will expand in diameter and possibly split.
    c. Nothing, because as ice expands, any increase in pressure causes an increase in temperature.
    d. The pipe will sag.

12. If a cubic foot of water at 60 degrees is heated to 212 degrees, how much will it weigh?
    a. 8.33 pounds
    b. 62.5 pounds
    c. 59.760 pounds
    d. 62.425 pounds

13. How many times its own volume will a cubic foot of water expand when heated to steam?
    a. 59.760 times   c. 212 times
    b. 62.5 times   d. 1728 times

14. How much will 30 gallons of water expand when heated from 60 degrees to 212 degrees?
    a. 1.154 gallons   c. 62.5 gallons
    b. 8.33 gallons   d. 212 gallons

15. What should be installed in every water heating boiler to allow for excess expansion?
    a. A gate valve   c. A pressure regulating valve
    b. A safety valve   d. A pressure shunt

UNIT 35 CONDUCTION, CONVECTION, AND RADIATION

OBJECTIVE

After studying this unit, the student will be able to:

- discuss conduction, convection, and radiation and their importance in the plumbing trade.

CONDUCTION

Conduction is the passage or transfer of heat from molecule to molecule, from the hottest to the coldest region of a substance. Every substance is made up of molecules which are in constant motion. When heat is added, molecules move at a faster rate. When a bar of iron is heated at one end, figure 35-1, the vibrations of the molecules increase. These vibrations are passed from molecule to molecule along the bar until the opposite end is hot.

The amount of heat lost is computed by the conductivity of the material. Figure 35-2 shows how heat in a room is lost through the ceiling, windows, and walls. Figure 35-3 shows how many Btu may be lost through a brick wall.

The rate at which a material conducts heat is expressed in the amount of British thermal units passing through 1 square foot of material 1 inch thick in 1 hour, with 1 degree temperature difference. Metal and glass have the highest ratings. Concrete, brick, and stone have a medium rating. Wood and the light, fluffy substances used for insulation have a low rating.

Copper and brass are used for heating coils because of their high rating, 240 Btu. Fiberboard, mineral wool, and asbestos are used for insulation because of a very low rating, .27 Btu.

CONVECTION

If part of a fluid body (liquid or gas) is heated, it expands and becomes lighter than

Fig. 35-1 Conduction through bar

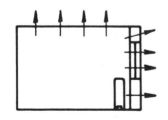

Fig. 35-2 Heat loss in a room

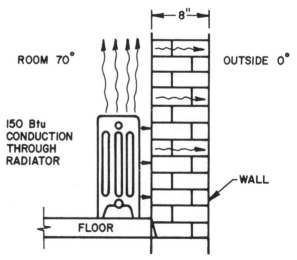

Fig. 35-3 Heat loss through brick wall (3 Btu)

the other parts. Since the molecules are free to move, the cooler, heavier part of the substance sinks to the bottom and pushes the hotter part of the substance to the top. Thus, ascending and descending currents are set in motion whenever any part of the liquid or gas at the bottom is heated to a hotter temperature than the rest. *Convection*, or circulation, is the circular movement caused by a difference in weight as a result of a temperature change.

If a full pan of cold water is heated, some of the water will run over the sides due to expansion. The pan will still be full of hot water, but lighter in weight. A cubic foot of water at 39.2 degrees weighs 62.425 pounds, while at 160 degrees it weighs 60.9 pounds, or 1.5 pounds less.

If a cubic foot of water at these temperatures is placed on opposite sides of a balance scale, the cold water will sink and raise the cubic foot of hot water. This takes place in all bodies of water when heat is applied.

Circulation between a heater and hot water tank takes place in the circulating pipes. The hot water travels up the flow (top pipe), and the cold water travels down the return pipe (bottom), figure 35-4. All circulating pipes should pitch in the direction that the water naturally travels to assist the flow.

Convection currents are also set in motion in air. In a fireplace, for instance, the air passing through the fire is heated and rises up the chimney, carrying the smoke and creating a draft. Convection can apply to a circular movement caused by differences in temperature in a body of either water or air.

RADIATION

Radiation is the passage of heat through space. The distance may be great or small. The heat absorbed is inversely proportional

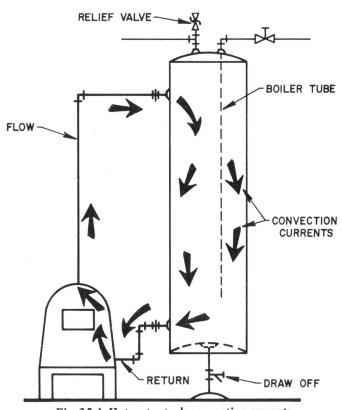

Fig. 35-4 Hot water tank convection currents

to the distance between the two objects. The heat absorbed from a radiator 5 feet away is 5 × 5 = 25, or 1/25th of the heat given off.

Heat from the sun, ninety-three million miles away, is absorbed by the earth. This heat travels through the infrared light in the atmosphere, not by the air itself. A 3/4-inch space of still air is a good insulator. Heat can travel through a vacuum; otherwise, we would receive no heat from the sun.

Heat loss by radiation is continually taking place wherever different temperatures exist. The sun heats the earth by day, and the earth's heat is given up to the air and objects at night. Heat produced in buildings in winter is lost to the outside air by conduction through the walls, glass, and roof. A dull, rough surface absorbs and radiates heat to a greater extent than a bright and smooth surface. This is because bright, shiny surfaces reflect heat. Aluminum foil is used for insulation because it reflects, or turns back, heat.

If a black cloth and a white cloth are placed on snow, the heat of the sun is absorbed to a greater extent by the black cloth and causes the snow to melt faster in this place.

REVIEW QUESTIONS

Multiple Choice

Select the best answer for each question.

1. What effect does heat have on the molecules in an iron bar?

   a. They vibrate faster.
   b. They turn a bluish color.
   c. They travel toward the cool end of the bar.
   d. They melt.

2. In what direction does the heat travel in an iron bar?

   a. Up
   b. Toward the cooler end
   c. Toward the heat
   d. North

3. Why are copper coils used in some gas water heaters?

   a. Copper is malleable.
   b. Copper is relatively inexpensive.
   c. Copper resists temperature changes.
   d. Copper has a high conductivity for heat.

4. Why is fiberglass insulation used to cover a steam main?

   a. Fiberglass insulation has a high Btu rating.
   b. Fiberglass insulation has a low Btu rating.
   c. Fiberglass protects the steel pipe from corrosion.
   d. Fiberglass is soft.

5. What materials are used for home insulation?

   a. Spun glass, asbestos, mineral wool
   b. Sheathing plywood
   c. Facing brick
   d. Sheet aluminum

6. Heat from a fireplace is felt on your face. What kind of heat is it?

    a. Conduction
    b. Convection

    c. Radiation
    d. Residual

7. Why is there a heat loss from a room to an outside wall?

    a. The insulation is probably poor.
    b. Heat will always travel toward the outside.
    c. Heat will always travel toward the cooler temperature.
    d. None of the above

8. Why is there no heat loss to an adjoining room?

    a. The temperatures are probably the same.
    b. The door between the rooms is probably closed.
    c. Internal walls are better insulated.
    d. None of the above

9. Heat transferred by convection can be thought of as

    a. a heated iron bar.
    b. the sun on a cold day.

    c. infiltration.
    d. currents in a gas or liquid.

10. Heat transferred by conduction can be thought of as

    a. a heated iron bar.
    b. the sun on a cold day.

    c. infiltration.
    d. currents in a gas or a liquid.

11. Circulation takes place in water because

    a. water becomes lighter as it gives up Btu.
    b. molecules vibrate when they are heated.
    c. warm water is lighter than cold water.
    d. of surrounding air temperatures.

12. In a container, why is the coolest water found at the bottom?

    a. Because of the shape of the container
    b. Because cold water, being heavier, sinks to the bottom
    c. Because radiation warms the water near the top
    d. Because the bottom is shaded

13. Why will snow under a black cloth melt faster than snow under a white cloth?

    a. Black absorbs heat, white reflects it.
    b. Black cloth must be made thicker and is therefore warmer.
    c. White cloth absorbs heat, black reflects it.
    d. none of the above

14. What causes a draft in a chimney?

    a. The wind blowing over the top
    b. Heat which is radiated from the chimney bricks
    c. Warm air rising in the flue
    d. Positive pressure within the house

15. What portion of heat is received at a distance of 5 feet from the radiator?

a. 1/2
b. 1/4
c. 1/5
d. 1/25 *

UNIT 36 AUTOMATIC STORAGE
GAS WATER HEATERS

OBJECTIVE

After studying this unit, the student will be able to:

• describe the construction and use of the automatic gas water heater.

WATER HEATERS

The automatic storage gas water heater is a vertical storage tank enclosed in a sheet metal case which is insulated to reduce heat loss, figure 36-1. One type has a preheater screwed into the bottom of the tank. The gas flame is directed against this preheater. The bottom of the tank and the flue also absorb heat.

The gas is automatically controlled by a thermostat. This thermostat is placed in the side of the tank so that the incoming cold water turns on the gas. When the water is heated to 140 or 160 degrees (the thermostat setting), it lowers the gas flame. This type of water heater has a graduated thermostat so the gas can be reduced gradually to a small flame. The flame maintains the heat of the water and acts as a pilot for the next operation.

A draw-off cock is installed at the bottom. A draft hood is placed on the flue pipe to prevent down draft which may put

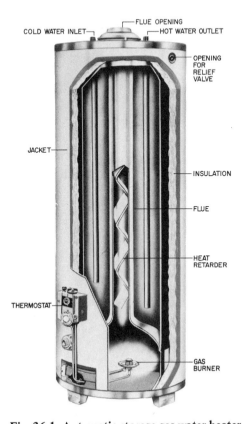

Fig. 36-1 Automatic storage gas water heater

out the flame. The flue is connected to the chimney so gas fumes can escape in case the pilot light goes out. Some of these heaters are equipped with Bunsen burners with fixed openings; others have adjustable air shutters.

Another type of gas water heater has no preheater but is arranged so the whole tank is a heating surface. The tank and the outer jacket are separated about 3/4 inch. This space is used as the flue.

Automatic gas water heaters are insulated with 1 1/2 inches of asbestos, mineral wool, or spun glass to conserve heat. Some heaters have a heat trap to prevent hot water from circulating in the supply line.

Every water-heating device must have a reliable relief valve to prevent explosions in case the thermostat fails to shut off the gas.

The heaters are equipped with an automatic pilot control. If the pilot goes out, the gas supply will not resume until the pilot is lit.

REVIEW QUESTIONS

Multiple Choice

Select the best answer for each question.

1. Hot water is taken from the water heater tank at
   a. a point 1/4 from the top.
   b. the base of the tank.
   c. the thermostat junction.
   d. the top.

2. Gas is automatically controlled by a
   a. hand valve.
   b. relief valve.
   c. draft hood.
   d. thermostat.

3. The draft hood prevents the
   a. tank from discoloring.
   b. flame from extinguishing.
   c. water from overheating.
   d. none of the above.

4. The flue pipe goes to the
   a. chimney.
   b. thermostat.
   c. automatic gas valve.
   d. relief valve.

5. If the pilot light goes out
   a. the heater may overheat.
   b. it stays out until it is relighted.
   c. it will come on again with the next heating cycle.
   d. it must be repaired.

6. What prevents the tank from exploding if the thermostat fails to shut off the gas?
   a. The pilot control
   b. The draft hood
   c. The relief valve
   d. The temperature regulator

7. A water heater may be insulated with
   a. copper flash coating.
   b. spun glass.
   c. rubber.
   d. particleboard.

8. The thermostat is placed near

    a. the flue pipe.
    b. the incoming cold water.

    c. the warmest part of the tank. °
    d. cellar door.

9. The draw-off cock is located near the

    a. hot water outlet.
    b. bottom of the tank. »

    c. pressure relief valve.
    d. flue pipe.

10. In some heaters, the _____ prevents hot water from circulating in the supply line.

    a. thermostat
    b. flue pipe ◦

    c. insulation
    d. heat trap

UNIT 37 THERMOSTATS

OBJECTIVES

After studying this unit, the student will be able to:

- describe the purpose of a thermostat.
- operate a thermostat.

THERMOSTATS

The gas valve of an automatic gas water heater is controlled by a thermostatic element inserted into the tank. The element, known as a *thermostat*, consists of a carbon rod encased in a copper tube which is closed at the end.

When the water in the tank is hot, the copper tube expands and permits the spring to close the gas valve. As hot water is drawn, cold water flows around the copper tube and causes the tube to contract. This pushes the carbon rod against the valve stem, lifting the valve from the seat. The gas then flows to the burner. As water heats around the copper tube, the tube again expands and the spring closes the gas valve. The length of the carbon rod and the copper tube must be exact, figure 37-1.

Most thermostats are equipped with a temperature indicator to regulate the water temperature.

Throttling thermostats are made so that the position of the valve seat is directly related to the temperature. Therefore, when the tank is cold, the valve is completely open. As the water warms, the valve gradually reduces the flow of gas until it is fully closed. Due to the slow opening, a bypass permits enough gas to maintain a minimum flame. This flame maintains the tank temperature and acts as a pilot light.

A bimetal element, called a *thermocouple*, is mounted in the pilot flame. If the pilot light goes out, a small electrical current generated by the thermocouple stops. This shuts off the gas. In this way unburned gas is not released into the living space.

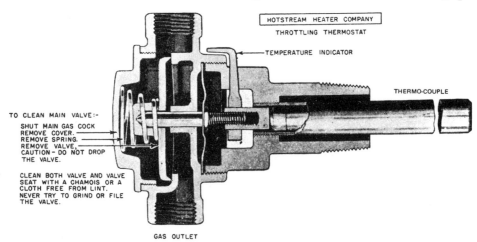

Fig. 37-1

Without a pilot light the heater, and perhaps the room or basement, may fill with gas. The gas could cause asphyxiation, or a spark may cause an explosion. Before the main gas burner can be relighted, the pilot must be lighted to heat the bimetal element.

DANGER: When lighting a gas water heater, hold a lighted match or taper to the burner before the gas is turned on. If it does not light, wait *five minutes* for the gas to leave the heater. Failure to do this may result in an explosion.

REVIEW QUESTIONS

Multiple Choice

Select the best answer for each question.

1. A carbon rod encased in a copper tube may be called a

   a. thermostat.   c. thermocouple.
   b. gas valve.   d. thermopile.

2. When the tank is cold,

   a. the copper tube expands and the valve opens.
   b. the copper tube expands and the valve closes.
   c. the copper tube contracts and the valve opens.
   d. the copper tube contracts and the valve closes.

3. When the water in the tank is hot,

   a. the carbon rod holds the valve open.
   b. the carbon rod moves out with the expanding copper tube and the valve closes.
   c. the copper tube holds the carbon rod tightly against the valve stem.
   d. none of the above.

4. The main advantage of a thermostatically-controlled valve on a water heater is

   a. it reduces condensation.
   b. it makes water heating with gas more economical.
   c. there is no chance of a gas explosion.
   d. it eliminates having to light the heater each time hot water is needed.

5. What is the difference in the reaction of a carbon rod and a copper tube to heat changes?

   a. The copper tube expands upon cooling, and the carbon rod does not change in length.
   b. The copper tube expands upon heating, and the carbon rod contracts.
   c. The copper tube contracts upon cooling, and the carbon rod does not change in length.
   d. The copper tube contracts upon heating, and the carbon rod expands.

6. When cold water enters a tank, what happens to the carbon rod?

    a. It is pressed against the valve stem by the contracting copper tube.
    b. It begins to contract.
    c. It opens the pilot control bypass valve.
    d. It is released by the expanding copper tube.

7. How is gas supplied to the pilot flame?

    a. By the action of the carbon rod on the valve
    b. By a bypass tube
    c. By excess gas from the main burner
    d. By the thermocouple

8. What is the purpose of the thermocouple?

    a. To shut off gas to the main burner if the pilot flame goes out
    b. To measure the temperature of the water tank
    c. To maintain the pilot flame
    d. To measure the heat of the pilot flame

9. Why is the thermostat located under the cold water inlet?

    a. To prevent corrosion buildup
    b. To react more quickly to hot water use
    c. To keep the mechanism cool
    d. To prevent calcium buildup

10. What should be used to clean the valve and valve seat?

    a. A fine-toothed file
    b. Double-fine grinding compound
    c. A chamois or lint-free cloth
    d. A half-round file

UNIT 38 RELIEF VALVES

OBJECTIVE

After studying this unit, the student will be able to:

- describe the purposes and uses of relief valves.

RELIEF VALVES

A *safety* or *relief valve* is a device which is placed on a closed water, air, steam, or oil system to prevent damage from excessive pressure. A properly placed and adjusted relief valve prevents pressure from exceeding a safe level, thereby preventing explosions.

Small relief valves, 1/2 to 2 inches, are made of brass with threaded connections. Large safety valves, 2 1/2 inches and larger, are made of cast iron or steel with brass seats and working parts. Large valves have flanged ends.

LEVER AND WEIGHT RELIEF VALVE

The *lever and weight relief valve* uses the principle of the lever and weight to balance the pressure within a tank or system,

figure 38-1. In this illustration, F is the fulcrum, W is the weight, S is the stem, L is the lever, and P is the internal pressure. Numbers and notches on the lever show where to place the weight to hold the desired pressure.

This relief valve operates like a third-class lever. When the internal pressure exceeds that for which the valve is set, it overcomes the weight and raises the washer from the seat. This relieves the pressure through the outlet and the weight closes the valve.

After the correct position of the weight has been determined, a hole is drilled through the weight and the lever. It may then be fastened in a permanent position with a bolt. The end of the lever beyond the weight

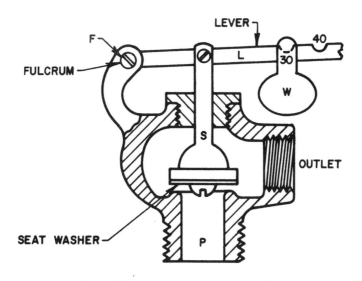

Fig. 38-1 Lever and weight relief valve

should be cut off. This prevents the movement of the weight beyond the safety point. Additional weight on the lever may cause an explosion.

Relief valves are set 25 percent above the water pressure on the system but below the working pressure of the equipment on which the valves are placed.

Relief valves are placed directly into or very close to the tanks or equipment which they protect. Valves should be opened occasionally to prevent closing by corrosion. The discharge of a relief valve should be run to 2 inches above an open floor drain to prevent contamination of the water supply by back siphonage.

Note: Check with the local code before using the lever relief valve.

SPRING RELIEF VALVES

The *spring relief valve* is sometimes known as a *pop valve* because it opens and closes suddenly. These valves use the tension of a coiled spring to withstand the internal pressure, figure 38-2. They are used on water piping, hot water tanks, pumps, heating boiler, air compressors, and oil lines. They are made in 1/2 and 3/4-inch sizes.

Pressure may be increased by water hammer. Water hammer is caused by the sudden closing of valves, by automatic switches failing to operate, or by the failure of oil burners or compressors to shut off.

Spring relief valves are set at about 25 percent higher than the pressure on the system. Adjustment is made by turning the adjusting screw in to increase the pressure. Turning it out decreases the discharge pressure.

A spring relief valve should have a large waterway to prevent clogging by corrosion. Some spring relief valves have a try handle attached which, if operated occasionally, prevents clogging. All parts exposed to water should be made of brass or bronze.

The adjusting screw must not be tightened. This would increase the discharge pressure to a dangerous point.

DIAPHRAGM RELIEF VALVES

The principle upon which the *diaphragm safety valve* operates is *area*. This means

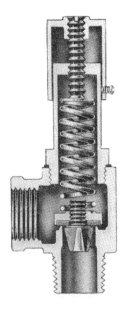

Fig. 38-2 Spring relief valve

that pressure is exerted equally in all directions upon equal area. The diaphragm is 35 times larger in area than the seat area. The heavy spring, figure 38-3(A), is necessary to overcome the difference in area.

The diaphragm relief valve is made in 1/2-inch size only. It has a 3/4-inch male thread inlet and a 1/2-inch outlet. The pressure ranges available are 50, 75, 100, 125, and 150 psi. Valves may be obtained which are set at any pressure between 5 to 160 psi.

Diaphragm relief valves are set with the spring chamber up. The valves are located as close to the hot water tank as possible. The distance from the hot water tank determines the effectiveness of a relief valve. The size of a relief valve is chosen according to its rated capacity in British thermal units

per hour. This capacity must be above the Btu input of the heater.

In the diaphragm relief valve in figure 38-3, the seat washer does not carry the heavy spring load. It is carried by the stops (B) in the valve body. The auxiliary spring (C) seats the valves with pressure when the excess pressure is relieved from the diaphragm. By using this lightweight spring, a soft seat washer may be used which opens with a slight increase of pressure.

CHOOSE THE PROPER RELIEF VALVE

Relief valves are seldom needed but must be effective when they are. Spring relief valves depend upon a heavy spring to close the valve against the pressure. After long periods of nonuse, the spring tension tends to

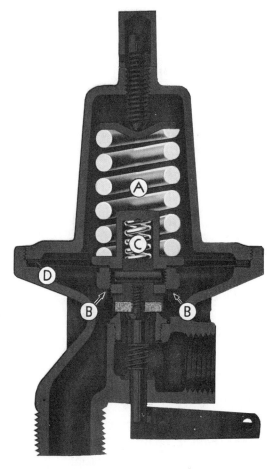

Fig. 38-3 Diaphragm relief valve

groove the seat washer. This fact, and the corrosion which takes place, causes the valve to stick. The pressure required to open it may damage the valve and create a dangerous situation.

Heated water can be as dangerous as dynamite. Therefore, every person installing heating devices should select the proper relief valve and know the best location for it on each particular job.

REVIEW QUESTIONS

Multiple Choice

Select the best answer for each question.

1. The purpose of a safety or relief valve is to

   a. prevent dangerous cross connections.
   b. prevent damage from excessive pressure.
   c. turn the energy source off in the event of pressure buildup.
   d. blow down the system.

2. Which of the following is not a safety valve?

   a. Lever and weight
   b. Regulating valve
   c. Diaphragm type
   d. Spring type

3. On the lever and weight relief valve, if the weight is moved out,

   a. the relief pressure increases.
   b. the relief pressure decreases.
   c. the system shuts off at the proper pressure.
   d. pressure escapes immediately.

4. The purpose of fixing the weight permanently to one spot on the lever is to

   a. prevent the valve from being damaged.
   b. abide by the National Plumbing Code.
   c. keep the weight from sliding off the end of the lever.
   d. none of the above

5. The lever and weight relief valve is

   a. adjusted for atmospheric pressure.
   b. mounted vertically.
   c. mounted only in a side outlet.
   d. checked with a gauge.

6. Why is the lever on a lever and weight relief valve cut off?

   a. To prevent unauthorized tampering with the pressure setting
   b. To reduce the weight to be lifted by internal pressure
   c. To prevent bounce in a blow-off situation
   d. To increase the pressure setting

7. A relief valve must be set

   a. above the working pressure of the equipment it serves.
   b. at least at 25 psi.
   c. above the minimum street pressure.
   d. at 25 percent above the water pressure.

8. On a piece of equipment with a rated working pressure of 125 psi and a maximum system pressure of 65 psi, the ideal setting for a lever relief valve would be

   a. 125 psi.   c. 65 psi.
   b. 81 psi.   d. twice the working pressure.

9. Valves should be opened occasionally by hand because

   a. the valve washers should be dampened occasionally.
   b. springs may suffer metal fatigue.
   c. corrosion may have made the valve inoperative.
   d. none of the above.

10. A pop valve is usually associated with the _____ relief valve.

    a. spring-type   c. diaphragm-type
    b. lever-type   d. fusible-type

11. The spring-type relief valve is usually made in sizes of

    a. 1/4 and 3/8 inch.   c. 1 and 1 1/4 inches.
    b. 1/2 and 3/4 inch.   d. larger than 2 inches.

12. On an adjustable-type relief valve, the pressure may be increased by

    a. selecting a larger size.
    b. crimping the cap.
    c. installing a thinner shim.
    d. turning the adjustment clockwise.

13. The purpose of the try handle is to

    a. make it easier to draw off water.
    b. identify the fitting as a relief valve.
    c. allow the fitting to be cleared out occasionally.
    d. test the boiler pressure.

14. Where should the discharge of a relief valve be piped?

    a. To an open floor drain if possible
    b. Into the return line in the heating boiler
    c. To a point below the flood rim of any fixture
    d. Into the expansion tank

15. Which of the following cannot force open a relief valve?

    a. Water hammer   c. Pumps which fail to shut off
    b. Excessive boiler pressure   d. High water table

16. What may happen if an automatic switch on a pump fails to shut off?

    a. The pump may cut off on the overload control.
    b. Pressure may build until the relief valve blows.
    c. The automatic bypass may open.
    d. None of the above

17. At what discharge pressure should a relief valve be set when the existing pressure is 40 psi?

    a. 40 psi   c. 60 psi
    b. 50 psi   d. 80 psi

18. What main advantage does the diaphragm relief valve have over others?

    a. It may be mounted in any position.
    b. It is corrosion resistant.
    c. It is sensitive to pressure changes.
    d. It is inexpensive.

19. What does water leaking around the adjusting screw in a diaphragm relief valve indicate?

    a. That the packing needs replacement
    b. That the screw needs tightening
    c. That the diaphragm is ruptured
    d. That the steam pressure is too high

Plumbing Sketch

Show by a sketch where a relief valve should be located on a hot water tank.

UNIT 39 ELECTRIC WATER HEATERS

OBJECTIVE

After studying this unit, the student will be able to:

- describe the electric water heater and how it operates.

ELECTRIC WATER HEATER

Electric water heaters, figure 39-1, are usually the storage type. Tank capacities are normally large enough to hold a 24-hour supply. They may be equipped with a time switch to take advantage of low electric rates at night. Electric water heaters are clean, safe, and present no danger of fire or odor.

In the electric water heater, the cold water is connected to the top of the storage tank. From that point it is delivered to the bottom of the storage tank through a dip tube. The warmer water rises to the top of the tank by circulation and is taken off at that point for distribution to the house. Most water heaters have an outlet in the side of the tank for a temperature/pressure relief valve. If there is no special opening for a relief valve, the valve is placed in a tee close to the hot water outlet in the tank.

An immersion-type heating element is shown on the heater in figure 39-1. It includes an upper and a lower element. The lower heating element maintains the standby

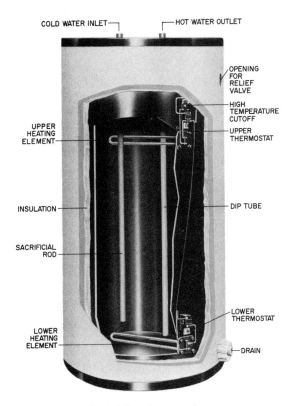

Fig. 39-1 Electric water heater

temperature of the tank. The upper element heats a small portion of the water to a higher temperature for immediate use.

Insulation of electric water heaters is very important. A bare tank at an average temperature of 130 degrees will lose 125 Btu per hour.

Water heaters can be obtained using either 120 or 240 volts of electrical current. 240-volt current is more efficient and economical to use. A magnesium rod is installed to retard corrosion. This is sometimes called a *sacrificial rod*.

REVIEW QUESTIONS

Multiple Choice

Select the best answer for each question.

1. The electric water heater should hold enough hot water to last
   a. 2 weeks.
   b. 24 hours.
   c. 12 hours.
   d. 2 hours.

2. The hottest water is found in
   a. the top of the tank.
   b. the relief valve.
   c. the bottom of the tank.
   d. the mid-heating cycle.

3. What is placed in the tank to slow corrosion?
   a. Heating element
   b. Relief valve
   c. Foam insulation
   d. Sacrificial rod

4. Why is the hot water taken off at the top of the tank?
   a. To draw in cold water
   b. To avoid turbulence within the tank
   c. To avoid setting off the relief valve
   d. Because that is where the hottest water is

5. The purpose of the time switch is to
   a. use electricity at the time when it is cheapest.
   b. shut off the water heater at night.
   c. avoid electrical shock when the heater is being repaired.
   d. limit the amount of electricity used.

6. The relief valve should be placed in
   a. the bottom of the tank.
   b. the cold water supply line.
   c. the outlet provided in the side of the tank.
   d. the gas line.

7. What type of heating element is used in electric water heaters?
   a. Immersion type
   b. Sacrificial anode
   c. High temperature
   d. Magnesium steel

8. Good insulation is necessary to prevent the loss of
   a. electricity.
   b. water.
   c. heat.
   d. none of the above.

9. What type of heaters use electricity most efficiently?

    a. Tall, thin heaters   c. Dark-colored heaters
    b. 240-volt heaters   d. USDA approved heaters

10. What maintains the standby temperature?

    a. Lower heating element
    b. Thermostatically-controlled valve
    c. Gas pilot light
    d. Upper heating element

UNIT 40 SUMMER-WINTER HOOKUPS

OBJECTIVE

After studying this unit, the student will be able to:

• describe the different types of summer-winter hookups.

SUMMER-WINTER HOOKUPS

The summer-winter hookup is a popular method of heating water in residential construction. A copper coil is immersed in the hot water of the house heating boiler. Circulating pipes connect the coil to the hot water storage tank, figure 40-1.

The water temperature in the heating boilers is maintained at 180 to 200 degrees by an aquastat which controls the oil burner. Water in the coil absorbs heat and circulates to the storage tank until it is the same temperature.

Since heated water rises and cold water sinks, it is best to use a horizontal storage tank. By placing the tank high, the cold water does not have to be lifted in the return pipe. If cold water has to be lifted, it slows circulation, figure 40-2. The cold water should be connected to the tank as shown in figure 40-1. If the cold water is connected to the return, it bypasses through the coil to the top of the tank where it chills the hot water, figure 40-3.

A weighted check (flow control) valve is placed on the flow main from the heating boiler to prevent hot water from flowing to the radiators until heat is required. This check valve is opened by the circulator which forces water into the radiators. One advantage of this system is that the heat maintained during the summer prevents the boiler from rusting. If the tank is insulated, fuel is saved and the heat is not noticed during the summer.

A three-way draw-off cock blows out the return in either direction. A relief valve should also be provided.

TANKLESS COIL

Many installations for small homes have a tankless coil. As its name implies,

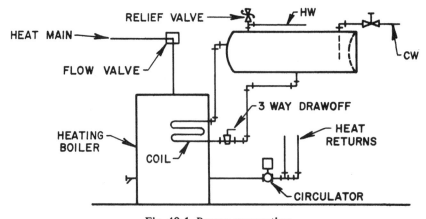

Fig. 40-1 Proper connection

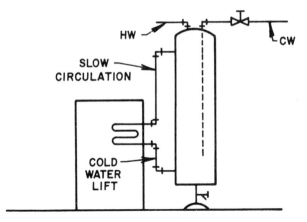

Fig. 40-2 Poor connection

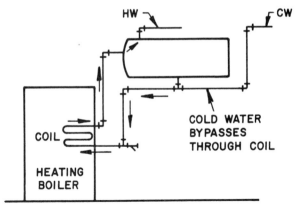

Fig. 40-3 Wrong connection

there is no storage tank. Enough hot water is supplied by immersing a large number of coils in the boiler water. Some installations have restrictions placed in the line to slow down the water flow through the coil. In this way, the water is kept in contact with the boiler water for a longer period.

A tempering valve is often used with the tankless coil system. The tempering valve mixes cold water with the outgoing hot water to maintain a preset temperature. In this way, a constant temperature is maintained. This eliminates the danger of the homeowner being scalded with hot water.

REVIEW QUESTIONS

Multiple Choice

Select the best answer for each question.

1. What is the advantage of the summer-winter hookup system?

   a. Boiler heat is maintained during the summer months so that rusting does not occur.
   b. The system is much less expensive than a water heater.
   c. A relief valve is unnecessary.
   d. It saves air-conditioning costs.

2. What turns the boiler on to maintain 180 to 200 degrees?

   a. The relief valve
   b. The tempering valve
   c. The aquastat
   d. The main breaker

3. The storage tank should be placed

   a. below the waterline of the boiler.
   b. outside of the house.
   c. vertically above the boiler.
   d. horizontally above the boiler.

4. What happens if the cold water is connected to the return line?

   a. The system is not affected.
   b. The cold water goes directly through the coil and reduces the temperature at the top of the tank.
   c. The coil is bypassed and the water is not heated.
   d. None of the above.

5. The purpose of the flow control valve is to

   a. restrict the flow of water to the tank.
   b. allow the return line to be blown out in either direction.
   c. keep hot water from going to the radiator in hot weather.
   d. separate air and water.

6. What is the name of the pipe which runs from the top of the heating coil to the tank?

   a. The relief circuit   c. The circulating bypass
   b. The return pipe   d. The flow pipe

7. What is an objection to using a vertical tank?

   a. The water will have a tendency to stratify.
   b. Circulation will be poor with the average basement height.
   c. The tank could be flooded with boiler water.
   d. There could be an excessive friction loss.

8. What is the purpose of the draw-off cock?

   a. To provide hot water to the basement
   b. To blow out the tank and lines occasionally
   c. To valve off the circulating lines
   d. To add boiler treatment chemicals

9. The purpose of the restriction in the tankless coil system is to

   a. keep it from being used on a job which is too big for the particular installation.
   b. eliminate the need for a flow control valve.
   c. slow down the flow so that it will pick up more heat.
   d. none of the above.

10. The tempering valve maintains a constant temperature by

   a. mixing varying amounts of cold water to the hot.
   b. adding a small amount of heat to water which is too cool.
   c. mixing boiler water to the domestic water to maintain a constant temperature.
   d. adding more water to the boiler.

UNIT 41 INDIRECT HEATERS
AND STEAM BOILERS

OBJECTIVE

After studying this unit, the student will be able to:

- describe the indirect heater and steam boiler.

INDIRECT HEATERS
AND STEAM BOILERS

Another method used to heat water is the *indirect heater*. This appliance consists of a copper coil within a cast-iron jacket. It

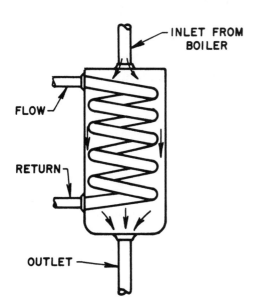

Fig. 41-1 Indirect heater

is piped so that it is below the waterline of the steam boiler. In this way, the hot water in the steam boiler circulates through the indirect heater and around the copper coil, figure 41-1.

A storage tank is used with the boiler and the indirect heater, figure 41-2. The flow and return from the copper coil is run to the storage tank. The water from the storage tank flows up through the coil and back to the tank. This is similar to the summer-winter hookup.

In the indirect heater, the outlet is at the bottom of the steam boiler. The inlet is run to the boiler outlets 2 inches below the waterline. In larger systems, all boiler sections are yoked together as in figure 41-2. A single connection may be made for the return. In figure 41-2 it should be noted that the four 1-inch connections are equal

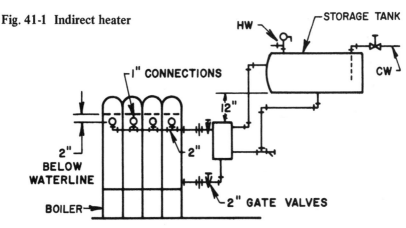

Fig. 41-2 Connections to steam boiler

to one 2-inch connection. It is advisable to place gate valves between the heater and the boiler so that repairs may be made.

The bottom of the storage tank should be at least 12 inches above the heater coil for gravity circulation. The upper connection on the tank hooks to the upper connection of the heater coil. The lower outlet of the tank hooks to the lower connection of the heater coil.

All water heating equipment must have a safety valve. Valves should never be placed between the heater and the tank.

The indirect heater may also be used with direct steam, figure 41-3. The indirect heater, in this case, is placed above the waterline of the boiler. The steam inlet is at the top; the return is taken from the

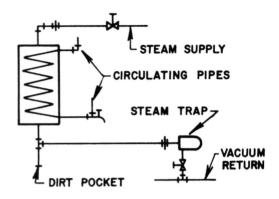

Fig. 41-3 Direct steam

bottom. A dirt pocket prevents dirt from entering the steam trap. Steam traps should be located at least 5 feet away to allow the steam condensation to cool before entering the trap. The trap outlet is connected to a vacuum return.

REVIEW QUESTIONS

Multiple Choice

Select the best answer for each question.

1. A copper coil encased in a cast-iron jacket describes what piece of water-heating equipment?

   a. A steam boiler   c. A storage tank
   b. An indirect heater   d. An expansion tank

2. The hot water from the steam boiler circulates

   a. upward.
   b. through the storage tank.
   c. inside the copper coils.
   d. through the indirect heater and around the copper coils.

3. How should the storage tank be positioned?

   a. 12 inches or more above the indirect heater
   b. Below the boiler waterline
   c. Below the coils
   d. In the attic

4. What size pipe equals the area of four 1-inch pipes?

   a. 1/2-inch pipe   c. 2-inch pipe
   b. 1 1/2-inch pipe   d. 4-inch pipe

5. What kind of valve is placed between the heater and the boiler?

    a. Relief valve
    b. Globe valve
    c. Gate valve
    d. Vacuum valve

6. The steam trap is located at least 5 feet away from the heater

    a. to give the condensation a chance to cool down.
    b. to lessen the chance of an explosion.
    c. to allow for fall.
    d. to protect the vacuum pump.

7. Where is the relief valve installed?

    a. Between the boiler and the indirect heater
    b. In the flow pipe from the indirect heater
    c. In the hot water line to the fixtures
    d. None of the above

Questions 8, 9, and 10 are based on figure 41-4.

8. The hot water from the steam boiler should be connected to which tapping of the indirect heater?

    a. 1   b. 2   c. 3   d. 4

9. The bottom tapping on the storage tank should be connected to which tapping of the indirect heater?

    a. 1   b. 2   c. 3   d. 4

10. Which outlet of the indirect heater returns to the boiler?

    a. 1   b. 2   c. 3   d. 4

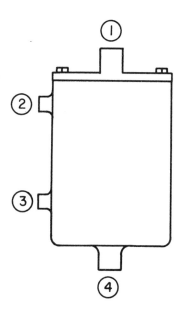

Fig. 41-4 Unit review illustration

UNIT 42 SOLAR WATER HEATING

OBJECTIVE

After studying this unit, the student will be able to:

- discuss how solar water heaters are constructed and used.

SOLAR WATER HEATING

Using the sun to heat water is not new. People in the south have used the sun for domestic water heating for many years. This method usually consists of a coil of copper soldered to a galvanized sheet and covered with glass, figure 42-1. The assembly is painted black and placed on the roof. The sun's rays penetrate the glass and warm the galvanized sheet and, along with it, the coiled copper tubing.

The heat from the sun's ray then accumulate under the glass cover. This is called the *greenhouse effect*. If a tank is placed in the peak of the roof, a natural circulation is set up between the collector and the tank, figure 42-2. The tank becomes warm and supplies hot water to the house. This is called a *thermosiphon system*.

Problems with the thermosiphon system occur when the collector becomes colder than the tank. A reverse circulation then takes place. This causes the heat already accumulated to be radiated back to the outside. It is also a difficult system to use where the temperature drops below freezing. However, thermosiphon systems are being tested in the north. Collectors are drained down during cold nights, and shutters are installed over the collector. Antifreeze solutions are used along with a heat exchanger. Most systems being manufactured use a pump and a heat exchanger to eliminate the problems of the thermosiphon system.

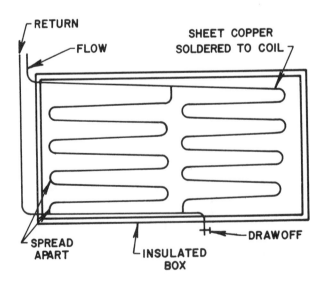

Fig. 42-1 Solar coil

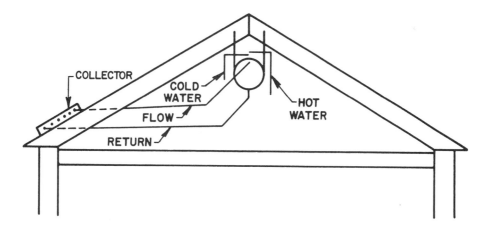

Fig. 42-2 Attic tank

COLLECTORS

The *collector* is the device which is exposed to the sun to collect the heat or Btu, figure 42-3. There are many different kinds of collectors. It may be a copper coil soldered to a copper sheet or aluminum sheets welded together with passages for liquid between them. It could also be a corrugated material where water trickles down the corrugations to pick up heat.

Because collectors are subjected to high heat, chemical reactions take place at a rapid rate. If these chemical reactions are not controlled, the collector will eventually develop leaks. Tap water contains percentages of free oxygen and minerals, and it often has a low pH (high acid) content. Therefore, additives are necessary to achieve a satisfactory life span of the unit. If antifreeze and anti-corrosion chemicals are added to the collector water, a heat-exchanger unit must be used.

Collectors should face toward the south, if possible, and never deviate more than 30 degrees from due south. The angle of the collector from the horizontal should be the latitude plus 10 degrees in sunny areas of the country, figure 42-4.

HEAT EXCHANGERS

If a coil of aluminum or copper tubing is immersed in a tub of cold water, and hot

Fig. 42-3 A solar collector

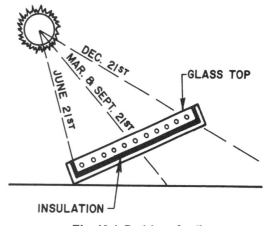

Fig. 42-4 Position of coil

water is passed through the coil, the tub of water will become warmer. This is the principle of the heat exchanger. The chemically-treated water from the collector is passed through the coil in the heat exchanger. The tap water in the exchanger tank is, in turn, heated by the coil.

There are, as in the case of the collector, many different heat-exchanger designs. Most manufacturers are combining the functions of the heat exchanger with that of the storage tank, figure 42-5.

THE STORAGE TANK

Because the sun does not shine during cloudy weather or at night, hot water must be stored for sunless periods. The size of the storage tank depends on the size of the collector. If the storage tank is too large, the collector may not have the capacity to bring the temperature up to a usable temperature (120 to 160 degrees).

A storage tank large enough to supply hot water under all possible conditions would have to be very large and very expensive. Because of this, solar-heating units are constructed to handle only ordinary weather conditions, and a conventional backup unit is added. Often, an electrical heating element is added to the storage tank. This heating element supplements the heat during long, sunless periods.

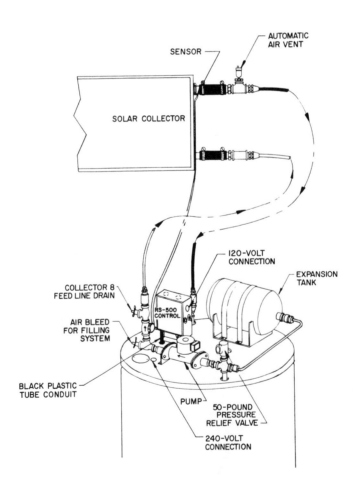

Fig. 42-5 Heat exchanger and storage tank

ACCESSORIES

When the storage tank is below the level of the collector, the collector fluid must be pumped up from the tank. A small-capacity pump must be part of the solar-heating equipment.

A thermostatically-controlled device controls the pump. It regulates the pump so it will only operate when the water in the collector is warmer than the water in the storage tank. Because the control must measure temperatures from two different points, it is often called a *temperature differential control box*.

The water within the storage tank can become so hot that a tempering valve must be installed. This prevents the occupants from being scalded. The tempering valve mixes some cold water with the outgoing hot water when it is necessary.

The collector and heat-exchange coils form a closed system. Since water expands when heated, an expansion tank must be installed. This provides a place for the expanded water to go. All water-heating systems must have a temperature/pressure relief valve also.

THE FUTURE OF SOLAR HEATING

The rising cost of fuel oil for generating electricity has influenced many people to consider solar-water heating. In general, gas and oil-fired water heaters are more economical at the present time. A customer who uses an electrical water heater, however, could very well benefit from solar heating. With the rapidly increasing cost of heating fuels, solar water heating will become more economical in all cases in the future.

Even though sunlight is free, solar heating equipment is expensive. In addition, colder areas and areas where the sun does not shine all the time still require a backup system. This is usually a conventional water heating system which comes on as needed.

Solar water-heating equipment for a home can cost $2400. If the life expectancy of the equipment is 20 years, the system must show a savings of 10 dollars a month to simply pay for itself. In general, if water heating is done by electricity, the solar system will pay for itself in 6 to 7 years. At present prices, if water heating is done by gas or oil, it may take as long as 30 years.

REVIEW QUESTIONS

Multiple Choice

Select the best answer for each question.

1. A solar heater will perform best in which location?
   a. Vermont
   b. Florida
   c. Illinois
   d. Pennsylvania

2. Another name for a natural circulation system is a
   a. pump-down system.
   b. heat-exchanger system.
   c. thermosiphon system.
   d. passive.

3. Glass is placed on the collector to create a
   a. greenhouse effect.
   b. natural circulation.
   c. proper angle.
   d. refraction index.

4. In thermosiphon systems, the flow may _____ after dark.
   a. leak
   b. stop
   c. continue
   d. reverse

5. The collector is placed

    a. on the roof.   c. in the cellar.
    b. in an unused room.   d. below ground level.

6. The purpose of the heat exchanger is to

    a. heat the water.
    b. transfer heat from the collector fluid to the domestic supply.
    c. prevent freezing of the domestic water supply.
    d. none of the above.

7. If the latitude is 34 degrees, the collector angle should be approximately

    a. 24 degrees from the vertical.
    b. 24 degrees from the horizontal.
    c. 44 degrees from the vertical.
    d. 44 degrees from the horizontal.

8. The temperature differential control

    a. allows the pump to run when the collector water is warm enough.
    b. turns the pump on when the exchanger water becomes excessively warm.
    c. keeps the water moving when the outside temperature drops below freezing.
    d. turns on the backup heating system.

9. Every water-heating system must have

    a. an alarm.   c. a relief valve.
    b. a room thermostat.   d. a pump.

10. The purpose of a tempering valve is to

    a. add enough cold water to the hot water to achieve a preset temperature.
    b. prevent metal fatigue within the system's components.
    c. relieve excessive pressure.
    d. add a given amount of antifreeze.

UNIT 43 FIXTURE INSTALLATION

OBJECTIVE

After studying this unit, the student will be able to:

- rough in and install fixtures following the manufacturer's instructions.

FIXTURE INSTALLATION

An important part of many plumbing jobs is the installation of the terminal equipment. These fixtures include toilets, lavatories, sinks, bathtubs, water softeners, air-conditioning units, water pumps, or heating equipment.

Fixtures are constantly changing in design. The plumber often has to rough in and install equipment that is unfamiliar. The plumber can do the job properly by following the manufacturer's rough-in information provided with each fixture.

A neat and clean fixture installation depends on an accurate rough-in. This involves locating and installing all pipes connected to the fixture. A careful examination of the manufacturer's information sheet will save time and material during rough-in and installation.

Fixtures and other equipment are often delicate and easily damaged. They should be protected after the installation is complete. There are a number of preparations on the market designed to protect fixture surfaces. Equipment can often be protected with its own packing material.

Pipe openings must be sealed after rough-in to prevent the entry of building residue. Pipes sticking out from walls and floors should be capped or plugged. Strainers should be removed and a piece of paper placed under them to prevent the entry of plaster and other building materials.

ROUGH-IN

The plumber must know wall and floor thicknesses before the fixtures are roughed in. Much of the information given on rough-in sheets is from the finished wall or floor. The plumber must allow for this distance. Backing boards and pipe support must be provided before subfloors and wall coverings are in place. *Stub pieces*, or capped pipe nipples, should penetrate the walls and floors. These help locate supply and drain lines after the floor and wall coverings are in place, figure 43-1.

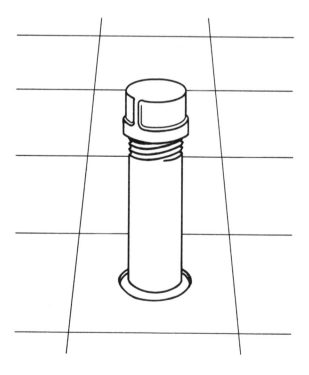

Fig. 43-1 A stub piece

YORKVILLE TOILET

VITREOUS CHINA

2128.023
2130.078

2128.023 ROUND FRONT   2130.078 ELONGATED
SHOWN WITH
3405.016 SUPPLY PIPE

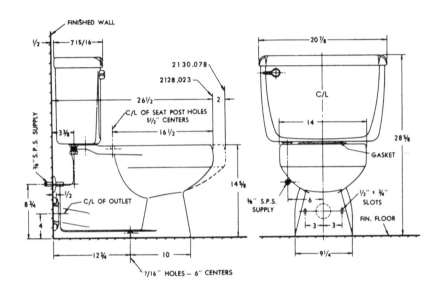

NOTE: ⅜" supply pipe not included with toilet and must be ordered separately.

IMPORTANT: Dimensions of fixtures are nominal and may vary within the range of tolerances
established by ANSI Standards A112.19.2.
These measurements are subject to change or cancellation. No responsibility is
assumed for use of superseded or voided leaflets.

Fig. 43-2 Toilet

Often, the plumber is just given approximate locations for fixtures. A good deal of judgment must be exercised in this case. Fixtures along a wall look best if the spaces between them are equal. However, allowable trap-arm lengths, an existing medicine cabinet, windows, and the swing of the door must be considered. There must be access to the bathtub plumbing and enough clearance to make the fixtures usable.

TOILETS

The toilet or water closet comes in a variety of styles. Wall-hung types require a special carrier behind the wall covering to hold the fixture weight. Some flush-tank toilets attach to the wall. These require careful alignment between the tank and the bowl in order to avoid leaks in the connecting flush ell.

There are also two-piece and one-piece close-coupled toilets. Floor-mounted toilets are roughed in 10, 12, or 14 inches from the finished wall, figure 43-2. The most common rough-in measurement is 12 inches. Assembly of the tank to the bowl must be done carefully to avoid breaking the fixture. Toilets are set on wax to the closet flange. Putty is not acceptable for setting the toilet bowl as it will dry out and shrink in time.

LAVATORIES

Lavatories or basins must be level and the proper flood-rim height above the floor, figure 43-3. A good quality, properly installed backing board or wall carrier is important.

Consult the manufacturer's rough-in sheet to determine the distance from the floor to the centerline of the screws in the

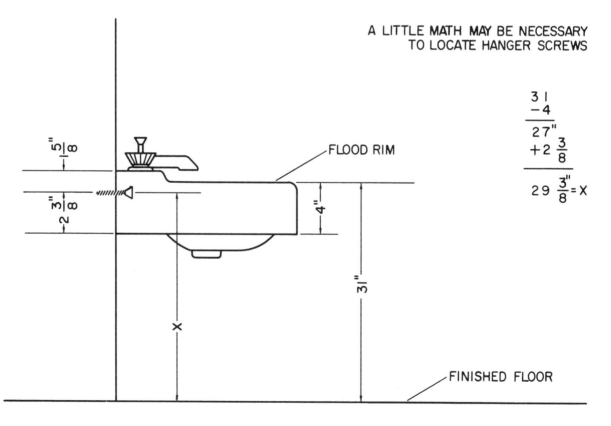

Fig. 43-3 Placing the lavatory

PENLYN LAVATORY

VITREOUS CHINA — WALL HUNG
SHOWN WITH
2248. SER. FITTING — 2303. SER. SUPPLIES
4401. SER. "P" TRAP

0372.029
(POP-UP DRAIN)

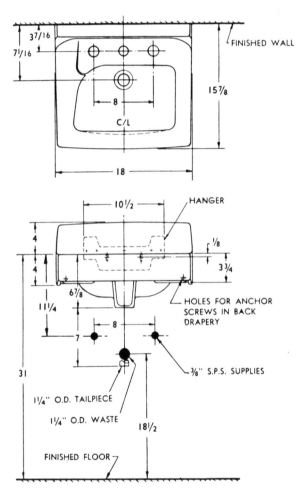

NOTE: FITTINGS NOT INCLUDED WITH FIXTURE AND MUST BE ORDERED SEPARATELY.

PLUMBER NOTE — Provide suitable reinforcement for all wall supports.

IMPORTANT: Dimensions of fixtures are nominal and may vary within the range of tolerances established by ANSI Standards A112.19.2.

These measurements are subject to change or cancellation. No responsibility is assumed for use of superseded or voided leaflets.

Fig. 43-4 Lavatory

hanger bracket, figure 43-4. Add the thickness of the finished floor. Using a 2-foot level, draw a horizontal line on the tile. Hold the hanger in place. Mark the screw holes. Drill the holes and set the hanger carefully. If a P trap is used, the centerline of the basin should be directly over the drain in the wall, figure 43-5.

BATHTUBS

Bathtubs come in a variety of shapes and sizes. They are made of cast iron, steel, fiberglass, or even wood. In most cases bathtubs are built in. They are installed against the studs in the wall rather than the finished wall.

The centerline of the drain hole is not usually in the center of the width of the tub, figure 43-6, page 180.

A left-hand tub has the waste opening on the left side when the viewer is standing at the skirted, long side of the tub. The drain line from the tub never comes up under the hole in the bottom of the bathtub. Use the rough-in sheet to locate the drain line.

The assembly which goes directly against the bathtub is called the *waste and overflow*. It may also be called a *trip lever assembly*. This assembly must be obtained separately, figure 43-7, page 181.

FAUCETS

Sink and basin faucets are often called *centersets*, figure 43-8, page 181. When installing, seal the faucet to the fixture with putty if a rubber gasket is not included. Sinks and lavatories (basins) installed in countertops must also be sealed to prevent water from leaking under the fixture.

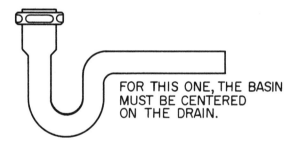

FOR THIS ONE, THE BASIN MUST BE CENTERED ON THE DRAIN.

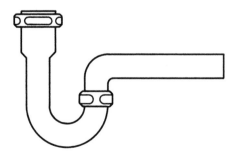

Fig. 43-5 Two kinds of P traps

SPECTRA BATH

ENAMELED CAST IRON — RECESS
SHOWN WITH
1303. SER. B/S SUPPLY FITTING &
1560. THRU 1562. SER. C.D.&O.

2605. SER.
2607. SER.

2607. SER. LEFT HAND OUTLET (SHOWN)
2605. SER. RIGHT HAND OUTLET (REVERSE DIMENSIONS)

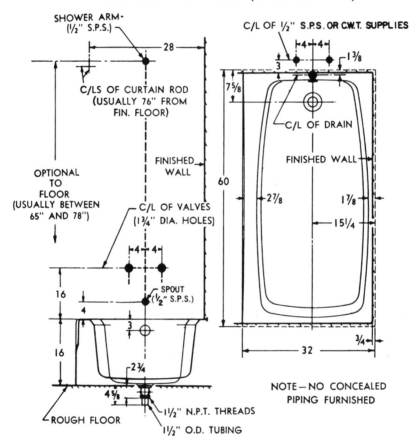

NOTE — NO CONCEALED
PIPING FURNISHED

NOTE: FITTINGS NOT-INCLUDED WITH FIXTURE AND MUST BE ORDERED SEPARATELY.

PLUMBER NOTE — Provide suitable reinforcement for all wall supports.

IMPORTANT: Dimensions of fixtures are nominal and may vary within the range of tolerances established by ANSI Standards A112.19.1.

These measurements are subject to change or cancellation. No responsibility is assumed for use of superseded or voided leaflets.

AMERICAN STANDARD

Fig. 43-6 Bathtub

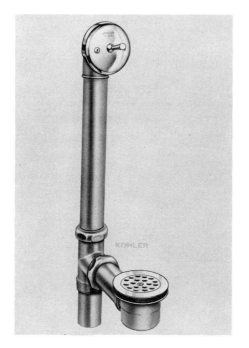

Fig. 43-7 Trip lever assembly

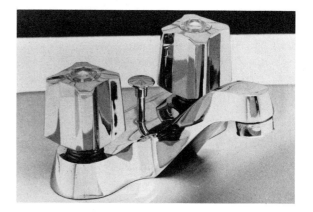

Fig. 43-8 Centerset

REVIEW QUESTIONS

Multiple Choice

Select the best answer for each question

1. The term to *rough in* means

   a. to allow for floor and wall coverings.
   b. to locate and install pipes before fixture installation.
   c. to install fixtures.
   d. figure out the job.

2. Stub pieces are installed

   a. to support fixtures.
   b. to protect fixtures and equipment.
   c. to support pipe.
   d. to locate pipes after wall and floor coverings are applied.

3. Fixtures should be located with _____ between them.

   a. 14 inches
   b. large spaces
   c. towel bars
   d. equal spaces

4. Care must be taken to avoid _____ the toilet bowl.

   a. aligning
   b. breaking
   c. twisting
   d. canting

5. In figure 43-2, how much room does the toilet take up from the wall to the front of the bowl?

   a. 8 7/16 inches
   b. 14 1/8 inches
   c. 26 1/2 inches
   d. 28 5/8 inches

6. In figure 43-3, how high is the flood rim of the lavatory above the finished floor?

   a. 18 1/2 inches
   b. 21 1/2 inches
   c. 31 inches
   d. 35 inches

7. For the lavatory shown in figure 43-4, how high above the floor should the drain line be roughed in?

   a. 6 7/8 inches
   b. 7 inches
   c. 19 3/4 inches
   d. None of the above

8. The centerline for the outlet of the bathtub shown in figure 43-6 is _____ from the long back edge.

   a. 1 3/8 inches
   b. 13 inches
   c. 15 1/4 inches
   d. 28 inches

9. In figure 43-6, the centerline of the waste line is roughed in _____ from the head end of the bathtub.

   a. 1 3/8 inches
   b. 7 5/8 inches
   c. 9 inches
   d. 52 3/8 inches

10. The waste and overflow is connected to the

    a. ball cock.
    b. sink trim.
    c. bathtub.
    d. shower diverter.

Project

Study the dimensions of the bathroom shown in figure 43-9 and the rough-in drawings in this unit. Design a bathroom to scale.

Use:

- Equal spaces where possible
- Adequate body clearances around fixtures
- Allowance for the swing of the door
- A partition for the exposed end of the bathtub if necessary

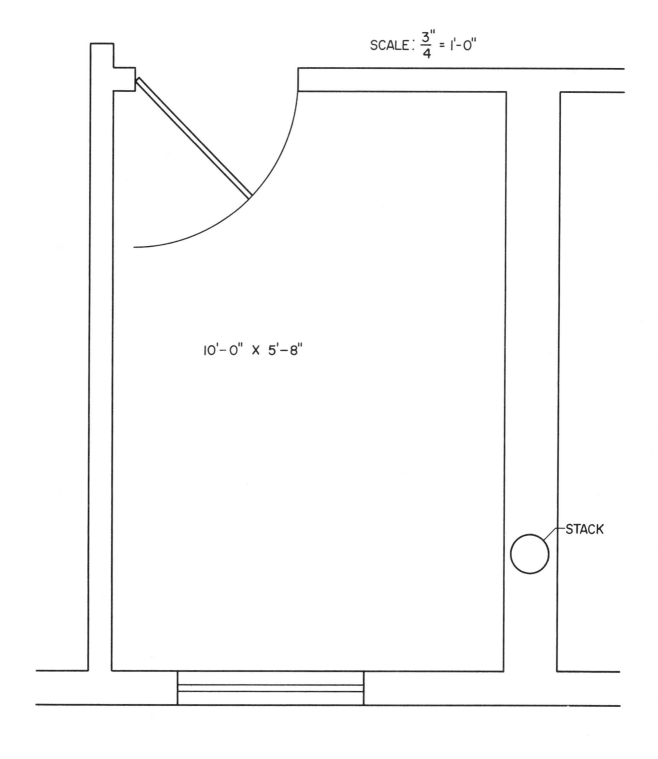

SCALE: $\frac{3''}{4} = 1'\text{-}0''$

10'- 0" X 5'- 8"

STACK

Fig. 43-9 Bathroom layout for assignment

WATER PRESSURE TO FEET HEAD

POUNDS PER SQUARE INCH	FEET HEAD	POUNDS PER SQUARE INCH	FEET HEAD
1	2.31	100	230.90
2	4.62	110	253.98
3	6.93	120	277.07
4	9.24	130	300.16
5	11.54	140	323.25
6	13.85	150	346.34
7	16.16	160	369.43
8	18.47	170	392.52
9	20.78	180	415.61
10	23.09	200	461.78
15	34.63	250	577.24
20	46.18	300	692.69
25	57.72	350	808.13
30	69.27	400	922.58
40	92.36	500	1154.48
50	115.45	600	1385.39
60	138.54	700	1616.30
70	161.63	800	1847.20
80	184.72	900	2078.10
90	207.81	1000	2309.00

NOTE: One pound of pressure per square inch of water equals 2.309 feet of water at 62° Fahrenheit. Therefore, to find the feet head of water for any pressure not given in the table above, multiply the pressure pounds per square inch by 2.309.

FEET HEAD OF WATER TO PSI

FEET HEAD	POUNDS PER SQUARE INCH	FEET HEAD	POUNDS PER SQUARE INCH
1	.43	100	43.31
2	.87	110	47.64
3	1.30	120	51.97
4	1.73	130	56.30
5	2.17	140	60.63
6	2.60	150	64.96
7	3.03	160	69.29
8	3.46	170	73.63
9	3.90	180	77.96
10	4.33	200	86.62
15	6.50	250	108.27
20	8.66	300	129.93
25	10.83	350	151.58
30	12.99	400	173.24
40	17.32	500	216.55
50	21.65	600	259.85
60	25.99	700	303.16
70	30.32	800	346.47
80	34.65	900	389.78
90	38.98	1000	433.00

NOTE: One foot of water at 62° Fahrenheit equals .433 pound pressure per square inch. To find the pressure per square inch for any feet head not given in the table above, multiply the feet head by .433.

BOILING POINTS OF WATER
AT VARIOUS PRESSURES

VACUUM, IN INCHES OF MERCURY	BOILING POINT	VACUUM, IN INCHES OF MERCURY	BOILING POINT
29	76.62	7	198.87
28	99.93	6	200.96
27	114.22	5	202.25
26	124.77	4	204.85
25	133.22	3	206.70
24	140.31	2	208.50
23	146.45	1	210.25
22	151.87	Gauge Lbs.	
21	156.75	0	212.
20	161.19	1	215.6
19	165.24	2	218.5
18	169.00	4	224.4
17	172.51	6	229.8
16	175.80	8	234.8
15	178.91	10	239.4
14	181.82	15	249.8
13	184.61	25	266.8
12	187.21	50	297.7
11	189.75	75	320.1
10	192.19	100	337.9
9	194.50	125	352.9
8	196.73	200	387.9

TAP AND DRILL SIZES
(American Standard Coarse)

SIZE OF DRILL	SIZE OF TAP	THREADS PER INCH	SIZE OF DRILL	SIZE OF TAP	THREADS PER INCH
7	1/4	20	49/64	7/8	9
F	5/16	18	53/64	15/16	9
5/16	3/8	16	7/8	1	8
U	7/16	14	63/64	1 1/8	7
27/64	1/2	13	1 7/64	1 1/4	7
31/64	9/16	12	1 13/64	1 3/8	6
17/32	5/8	11	1 11/32	1 1/2	6
19/32	11/16	11	1 29/64	1 5/8	5 1/2
21/32	3/4	10	1 9/16	1 3/4	5
23/32	13/16	10	1 11/16	1 7/8	5
			1 25/32	2	4 1/2

MINUTES CONVERTED TO DECIMALS OF A DEGREE

MIN.	DEG.	MIN.	DEG.	MIN.	DEG.	MIN.	DEG.	MIN.	DEG.	MIN.	DEG.
1	.0166	11	.1833	21	.3500	31	.5166	41	.6833	51	.8500
2	.0333	12	.2000	22	.3666	32	.5333	42	.7000	52	.8666
3	.0500	13	.2166	23	.3833	33	.5500	43	.7166	53	.8833
4	.0666	14	.2333	24	.4000	34	.5666	44	.7333	54	.9000
5	.0833	15	.2500	25	.4166	35	.5833	45	.7500	55	.9166
6	.1000	16	.2666	26	.4333	36	.6000	46	.7666	56	.9333
7	.1166	17	.2833	27	.4500	37	.6166	47	.7833	57	.9500
8	.1333	18	.3000	28	.4666	38	.6333	48	.8000	58	.9666
9	.1500	19	.3166	29	.4833	39	.6500	49	.8166	59	.9833
10	.1666	20	.3333	30	.5000	40	.6666	50	.8333	60	1.0000

DECIMAL EQUIVALENTS OF FRACTIONS

INCHES	DECIMAL OF AN INCH	INCHES	DECIMAL OF AN INCH
1/64	.015625	7/16	.4375
1/32	.03125	29/64	.453125
3/64	.046875	15/32	.46875
1/20	.05	31/64	.484375
1/16	.0625	1/2	.5
1/13	.0769	33/64	.515625
5/64	.078125	17/32	.53125
1/12	.0833	35/64	.546875
1/11	.0909	9/16	.5625
3/32	.09375	37/64	.578125
1/10	.10	19/32	.59375
7/64	.109375	39/64	.609375
1/9	.111	5/8	.625
1/8	.125	41/64	.640625
9/64	.140625	21/32	.65625
1/7	.1429	43/64	.671875
5/32	.15625	11/16	.6875
1/6	.1667	45/64	.703125
11/64	.171875	23/32	.71875
3/16	.1875	47/64	.734375
1/5	.2	3/4	.75
13/64	.203125	49/64	.765625
7/32	.21875	25/32	.78125
15/64	.234375	51/64	.796875
1/4	.25	13/16	.8125
17/64	.265625	53/64	.828125
9/32	.28125	27/32	.84375
19/64	.296875	55/64	.859375
5/16	.3125	7/8	.875
21/64	.328125	57/64	.890625
1/3	.333	29/32	.90625
11/32	.34375	59/64	.921875
23/64	.359375	15/16	.9375
3/8	.375	61/64	.953125
25/64	.390625	31/32	.96875
13/32	.40625	63/64	.984375
27/64	.421875	1	1.

STANDARD PIPE DATA

NOMINAL PIPE DIAM. IN INCHES	ACTUAL INSIDE DIAM. IN INCHES	ACTUAL OUTSIDE DIAM. IN INCHES	WEIGHT PER FOOT POUNDS	LENGTH IN FEET CONTAINING ONE CUBIC FOOT	GALLONS IN ONE LINEAL FOOT
1/8	.269	.405	.244	2526.000	.0030
1/4	.364	.540	.424	1383.800	.0054
3/8	.493	.675	.567	754.360	.0099
1/2	.622	.840	.850	473.910	.0158
3/4	.824	1.050	1.130	270.030	.0277
1	1.049	1.315	1.678	166.620	.0449
1 1/4	1.380	1.660	2.272	96.275	.0777
1 1/2	1.610	1.900	2.717	70.733	.1058
2	2.067	2.375	3.652	49.913	.1743
2 1/2	2.469	2.875	5.793	30.077	.2487
3	3.068	3.500	7.575	19.479	.3840
3 1/2	3.548	4.000	9.109	14.565	.5136
4	4.026	4.500	10.790	11.312	.6613
4 1/2	4.560	5.000	12.538	9.030	.8284
5	5.047	5.563	14.617	7.198	1.0393
6	6.065	6.625	18.974	4.984	1.5008
8	7.981	8.625	28.554	2.878	2.5988
10	10.020	10.750	40.483	1.826	4.0963

CONVERSION FACTORS

PRESSURE

1 lb. per sq. in.	=	2.31 ft. water at 60°F
	=	2.04 in. hg at 60°F
1 ft. water at 60°F	=	0.433 lb. per sq. in.
	=	0.884 in. hg at 60°F
1 in. Hg at 60°F	=	0.49 lb. per sq. in.
	=	1.13 ft. water at 60°F
lb. per sq. in. Absolute (psia)	=	lb. per sq. in. gauge (psig) + 14.7

TEMPERATURE

°C	=	(°F – 32) × 5/9

WEIGHT OF LIQUID

1 gal. (U.S.)	=	8.34 lb. × sp. gr.
1 cu. ft.	=	62.4 lb. × sp. gr.
1 lb.	=	0.12 U.S. gal. ÷ sp. gr.
	=	0.016 cu. ft. ÷ sp. gr.

FLOW

1 gpm	=	0.134 cu. ft. per min.
	=	500 lb. per hr. × sp. gr.
500 lb. per hr.	=	1 gpm ÷ sp. gr.
1 cu. ft. per min. (cfm)	=	448.8 gal. per hr. (gph)

WORK

1 Btu (mean)	=	778 ft. lb.
	=	0.293 watt hr.
	=	1/180 of heat required to change temp of 1 lb. water from 32°F to 212°F
1 hp-hr	=	2545 Btu (mean)
	=	0.746 Kwhr
1 Kwhr	=	3413 Btu (mean)
	=	1.34 hp-hr

POWER

1 Btu per hr.	=	0.293 watt
	=	12.96 ft. lb. per min.
	=	0.00039 hp
1 ton refrigeration (U.S.)	=	288,000 Btu per 24 hr.
	=	12,000 Btu per hr.
	=	200 Btu per min.
	=	83.33 lb. ice melted per hr. from and at $32°$ F.
	=	2000 lb. ice melted per 24 hr. from and at $32°$ F.
1 hp	=	550 ft. lb. per sec.
	=	746 watt
	=	2545 Btu per hr.
1 boiler hp	=	33,480 Btu per hr.
	=	34.5 lb. water evap. per hr. from and at $212°$ F.
	=	9.8 kw.
1 kw.	=	3413 Btu per hr.

MASS

1 lb. (avoir.)	=	16 oz. (avoir.)
	=	7000 grain
1 ton (short)	=	2000 lb.
1 ton (long)	=	2240 lb.

VOLUME

1 gal. (U.S.)	=	128 fl. oz. (U.S.)
	=	231 cu. in.
	=	0.833 gal. (Brit.)
1 cu. ft.	=	7.48 gal. (U.S.)

WEIGHT OF WATER

1 cu. ft. at $50°$ F. weighs 62.41 lb.

1 gal. at $50°$ F. weighs 8.34 lb.

1 cu. ft. of ice weighs 57.2 lb.

Water is at its greatest density at $39.2°$ F.

1 cu. ft. at $39.2°$ F. weighs 62.43 lb.

CONVERSION CONSTANTS

TO CHANGE	TO	MULTIPLY BY
Inches	Feet	0.0833
Inches	Millimeters	25.4
Feet	Inches	12
Feet	Yards	0.3333
Yards	Feet	3
Square inches	Square feet	0.00694
Square feet	Square inches	144
Square feet	Square yards	0.11111
Square yards	Square feet	9
Cubic inches	Cubic feet	0.00058
Cubic feet	Cubic inches	1728
Cubic feet	Cubic yards	0.03703
Cubic yards	Cubic feet	27
Cubic inches	Gallons	0.00433
Cubic feet	Gallons	7.48
Gallons	Cubic inches	231
Gallons	Cubic feet	0.1337
Gallons	Pounds of water	8.33
Pounds of water	Gallons	0.12004
Ounces	Pounds	0.0625
Pounds	Ounces	16
Inches of water	Pounds per square inch	0.0361
Inches of water	Inches of mercury	0.0735
Inches of water	Ounces per square inch	0.578
Inches of water	Pounds per square foot	5.2
Inches of mercury	Inches of water	13.6
Inches of mercury	Feet of water	1.1333
Inches of mercury	Pounds per square inch	0.4914
Ounces per square inch	Inches of mercury	0.127
Ounces per square inch	Inches of water	1.733
Pounds per square inch	Inches of water	27.72
Pounds per square inch	Feet of water	2.310
Pounds per square inch	Inches of mercury	2.04
Pounds per square inch	Atmospheres	0.0681
Feet of water	Pounds per square inch	0.434
Feet of water	Pounds per square foot	62.5
Feet of water	Inches of mercury	0.8824
Atmospheres	Pounds per square inch	14.696
Atmospheres	Inches of mercury	29.92
Atmospheres	Feet of water	34
Long tons	Pounds	2240
Short tons	Pounds	2000
Short tons	Long tons	0.89285

Acknowledgments

The authors extend special thanks to Albert W. Smith and Mario J. Fala for their technical assistance and review of ADVANCED PLUMBING.

Illustrations

American Standard — 43-2, 43-4, 43-6
Bradford-White — 36-1, 39-1
Copper Development Association, Inc. — 27-1, 27-2, 27-3, 27-4, 27-5
Kohler Company — 43-7, 43-8
Mor-Flo Industries Inc. — 42-5
National Standard Plumbing Code — 7-4, 20-3, 20-4, 20-5, 21-1, 21-2, 21-3, 23-2, 23-5
Photoworkshop — title page photo
Tube Turns — 5-5, 5-6, 5-7, 5-8, 5-9, 5-10, 5-11, appendix charts

Delmar Staff

Mark W. Huth, Sponsoring Editor
Kathleen E. Beiswenger, Associate Editor
Barbara A. Brown, Copy Editor